Agricultural Development and Productivity

Lessons from the Chilean Experience

Agricultural Development and Productivity

Lessons from the Chilean Experience

by
Pierre R. Crosson

Published for Resources for the Future, Inc.
by The Johns Hopkins Press, Baltimore and London

Pierre R. Crosson is a senior research associate with Resources for the Future. The manuscript was edited by Pauline Batchelder. Charts were drawn by Clare and Frank J. Ford. The index was prepared by Margaret Stanley.

RFF editors: Henry Jarrett, Vera W. Dodds, Nora E. Roots, Tadd Fisher.

Contents

List of Tables

CHAPTER 4

CHAPTER 5

CHAPTER 6

APPENDIX

Foreword

The role of agriculture in the economic development of less developed countries has received increasing attention in recent years, reflecting a growing awareness among economists and policy-makers that the problems of underdevelopment cannot be solved without a solution to the problem of agricultural stagnation. Achievement of a dynamic agriculture capable of generating steadily rising levels of productivity, therefore, has become a prime policy objective through the developing world. Our knowledge of how to reach this objective, however, still is quite limited. While there is an abundant and growing literature concerned with agricultural development, the body of empirical work necessary to test the validity of ideas and models found in the literature is inadequate.

The present study is a contribution toward making up the deficiency of empirical work. Chile is especially apt for this purpose because there is wide agreement that the pace of its agricultural development has fallen well short of its potential. Hence examination of the Chilean experience promises to yield some fruitful insights into the process of agricultural development, particularly regarding the obstacles to more rapid growth in productivity. Pierre Crosson has undertaken such an examination in this study. I believe he has succeeded in advancing our understanding of agricultural development, and that his study, therefore, will be of interest to the community of scholars, planners, and policy-makers concerned with that vital subject.

The book is one of RFF's studies dealing with Latin American problems. Other books in this series are: Nathaniel Wollman, *The Water Resources of Chile* (1968); Orris C. Herfindahl, *Natural Resource Information for Economic Development* (1969); Harvey S. Perloff, *Alliance for Progress* (1969); Joseph Grunwald and Philip Musgrove, *Natural Resources in Latin American Development* (1970); and Raymond F. Mikesell et al., *Foreign Investment in the Petroleum and Mineral Industries: Case Studies of Investor-Host Country Relations* (1970); all published by The Johns Hopkins Press.

JOSEPH L. FISHER
Resources for the Future, Inc.

Preface

There is general agreement that agricultural production in Chile for many years has fallen well short of its potential, and that both the great mass of farm people and the economy as a whole have suffered as a consequence. Blessed with a generally equable climate and rich soils, the great Central Valley of Chile has been described by Theodore Schultz as "probably the best piece of farm real estate in the world" outside California. Despite this first-rate natural endowment, Chilean agricultural production in recent decades has barely kept pace with population growth and has fallen well short of the growth in total demand. One important consequence is that scarce foreign exchange has had to be diverted from vitally important capital formation to the financing of imported foods, many of which could be economically produced in Chile if the agricultural sector were more efficiently organized.

This study attempts to make some progress in understanding why the performance of Chilean agriculture has fallen short of its potential. Two approaches to the problem are developed. One focuses on the overall performance of agriculture in the 1950's, making use of such nationwide data as are available but relying for the most part on data obtained from a sample of farms located in the Central Valley. The second approach involves analysis of total-productivity differences between groups of farms in the sample, attempting both to measure the magnitude of these differences and to account for them. These two approaches were adopted in recognition that the total performance of agriculture reflects the play of forces which impinge more or less evenly on all farmers (the first approach) and of those which have a differential impact (the second approach). Full understanding, therefore, requires examination of both sets of forces.

The decision to focus the bulk of the analysis on the 1950's was dictated primarily by the availability of data. However, some evidence is presented indicating that the situation prevailing in the 1950's persisted well into the

1960's. One advantage of limiting the treatment to the earlier decade is that performance was not seriously affected in that period by the land reform controversy that has dominated discussion of Chilean agriculture since 1964.

No claim is made that the analysis is exhaustive or that the conclusions reached are definitive. It is believed, however, that the study succeeds in illuminating some significant aspects of the performance of Chilean agriculture in the 1950's and early 1960's.

Of course, most of the results of the study are directly relevant only to Chile. However, the Chilean situation has much in common with that of other countries of Latin America, particularly with respect to land tenure conditions and government price, credit, and foreign-trade policies affecting agriculture. For these reasons it is believed that the study will be of interest to all students of Latin American agricultural development, as well as to policy makers attempting to cope with the difficult job of improving agricultural performance. It is with the latter particularly in mind that a chapter has been included dealing with planning implications of the analysis. This chapter stresses the methodology employed in analyzing the Chilean situation, attempting to show how this methodology can be useful to agricultural planners and policy makers wherever they may be. Throughout the discussion every effort is made to indicate the limitations of the methods used as well as their strong points. They are presented as valuable additions to the planner's tool kit of planning techniques, not as a set of master devices which render all others obsolete.

The study could not have been done without the assistance of many persons in Chile and had there not already been in existence a substantial body of highly professional research dealing with various aspects of the country's agricultural economy. The latter are cited appropriately throughout the study. I would like, however, to make a general acknowledgment here of the particularly valuable work done by the joint Economic Commission for Latin America/Food and Agriculture Organization group in Santiago on the use of fertilizers, pesticides, and machinery in Chile, and by the Economics Institute of the University of Chile and the Department of Agricultural Economics of the Catholic University of Chile on a variety of topics related to agriculture. In my view, the scope and quality of the work of these several institutions have not been adequately recognized. It is hoped that this study will help to compensate for this oversight.

Among the many individuals who have contributed to the study I must put Cornelio Marchán in first place. He was associated with me throughout the substantive work on the study, assisting not only in gathering and processing data but also in interviewing. In addition, he made a number of most valuable suggestions. In general, his contribution went well beyond that of a research assistant. Alden Gaete worked with me toward the end and made particularly valuable contributions to the design and field work connected with a survey of farmers, as did Atilio Giglio. Solon Barraclough, Michael Nelson, Delbert Fitchett, Wade Gregory, Louis Nixon, Henry Bruton, Sam Engene, Sterling Brubaker, and Carlos Plaza and Jorge Alcázar, both of the Latin American Institute for Economic and Social Planning, read and made valuable comments on preliminary versions of the study. Mr. Alcázar also reviewed the initial questionnaire used in the above-mentioned field survey and made a number of valuable suggestions for eliminating ambiguous or irrelevant questions. Paul Aldunate, professor at Chile's Catholic University, acted as consultant on some of the technical aspects of production function analysis and made many helpful suggestions. Irving Hoch, my colleague at Resources for the Future, also read these parts of the manuscript and made a number of most useful comments. If I have nonetheless gone astray on some part of this analysis, the fault is mine. Robert Steinberg, then of Resources for the Future, did the programming for the production function analysis and also did some very useful statistical testing.

Many persons submitted to interviews covering various aspects of the study. In this respect I wish to acknowledge my indebtedness to Eduardo García, Rolando Chateauneuf, Carlos Llona, Ambrosio García, Enrique Raymond, Sergio Cruz, Julio Bravo, N. E. Grimsditch, Abraham Dubinosky, Ricardo Edwards, and Julio Larenas.

A very special acknowledgment must go to the personnel of the Ministry of Agriculture's Agricultural and Livestock Production Division of the Agricultural and Livestock Service (División de Producción Agropecuaria, Servicio Agrícola y Ganadero), formerly the Department of Agricultural Economics. In particular, I must mention Santos Pérez, Director of the Agricultural and Livestock Production Division; Osvaldo Luco, Director, and Augusto Donoso, Sub-Director, of the Division's office in Rancagua; as well as Marcello Carvallo, Hernán Burgos, and Wolf Maige. They were repeatedly called upon to give their assistance and cooperation in making data available from an earlier survey and in conducting the one mentioned above. They gave unstintingly of both. It is not too much to say that without their help the study could not have been undertaken. One frequently hears that Latin American government officials are uncooperative or even unfriendly to research which touches their fields of operation. Nothing could have been further from this writer's experience.

When the study was in the final stages of editing I had the opportunity to return to Santiago to review much of the work that had been done in the Chilean Agricultural Planning Office (Oficina de Planificación Agrícola, ODEPA) in connection with the agricultural development plan for 1965–80 and to talk at length with senior personnel responsible for the plan. This was a most valuable opportunity, for which I wish to record here my thanks and appreciation. It should be noted, however, that the ODEPA personnel with whom I spoke are in basic disagreement with some aspects of my analysis, particularly that dealing with the interpretation of the economic behavior of large landowners. They most emphatically do not endorse this part of the analysis, nor are they necessarily in agreement with any other part of it.[1]

Finally, I must express my gratitude to Resources for the Future and the Latin American Institute for Economic and Social Planning for providing me the opportunity to undertake this study under conditions which could hardly have been better. I was domiciled in the latter institution for over two years and received all I asked in the way of hospitality and a congenial environment in which to work. For this, Cristóbal Lara, Assistant Director of the Institute, and Francis Shomaly, Chief Administrative Officer, are especially to be thanked. In addition, I had useful discussions at various stages in the work with a number of Institute staff members, notably Estevam Strauss, Norberto González, Osvaldo Sunkel, and Jesús González.

I am particularly grateful to Kelvin Scott, who gave a careful editorial review to an earlier draft of the entire manuscript.

As always, the author alone is responsible for any remaining errors of fact, logic, or common sense.

PIERRE CROSSON

March 1970

---

[1]The ODEPA position is set forth at length in República de Chile, Ministerio de Agricultura, *Plan de desarrollo agropecuario 1965-1980*, 4 vols. (Santiago, 1968).

Agricultural Development
and
Productivity

Lessons from the Chilean Experience

The Analytical Framework

Agricultural development involves improvement in the productivity of the resources employed by farmers. Of course, output can be expanded with greater employment of traditional land and labor inputs without any increase in productivity, and such increases would be welcomed in most less developed countries. However, they reflect merely the extension of existing primitive techniques of production to a wider land area. At best such output increments are achieved at constant cost. As the economy begins to impinge on available land resources, the extension of traditional techniques inevitably results in higher production costs. This can be avoided only by substituting new, more productive inputs for the traditional ones of land and labor. This process of input substitution is at the heart of agricultural development, and its pace determines the rate at which development occurs.

This view of the process of development is considerably more useful than that which attributes productivity gains to "technological change," since economists traditionally have viewed technology as outside the purview of economic analysis. To attribute productivity increases to "technological change," therefore, is to assert implicitly that economists have nothing to say, *qua* economists, about the process of development. This is not the case, however, when advancing productivity is viewed as the outcome of a process of input substitution. The reason is that the invention of new inputs and their introduction into the production process is importantly affected by considerations of cost and prospective gain, factors with which economics has traditionally been preeminently concerned.[1] But the old analytical tools alone will not suffice, because some of the new inputs of major importance respond to factors outside the range of traditional economics. Improvements in the ca-

[1] This way of viewing the sources of productivity growth owes much to T. W. Schultz, *Transforming Traditional Agriculture* (New Haven and London: Yale University Press, 1964), Ch. 9.

1

pacity of men to manage productive processes is surely one of the principal sources of rising productivity. Yet the expansion of human skills does not respond only or even primarily to the flux of market conditions typically studied by economists. The growth of scientific knowledge, a process as old as man himself, is above all a social process in the sense that it owes little to private calculations of cost and gain. In part it reflects something basic in the human animal, man's restless thrust to cope with the problems endlessly thrown up by powerful natural forces. The great advances in knowledge from the mastery of fire and the domestication of plants and animals to nuclear physics and the technology of space exploration surely owe much more to man's drive to understand and control his environment than to the market calculus.

The socialization of the expansion of basic knowledge is also explicable in part by factors which all economists readily understand. Basic research, at least nowadays, is expensive in manpower and frequently in equipment. But it is virtually impossible for the scientific pioneer to capture more than a small portion of the material return generated by his efforts. Instead it quickly becomes part of mankind's common stock of basic information available to anyone with the wit to use it. Consequently, private incentives to invest in basic scientific research are weak.[2]

The socialization of the "knowledge industry" would not present special problems for traditional economic analysis if the allocation of resources to and within this industry were responsive to relative prices of inputs and outputs. But nonmarket criteria are likely to be more important than prices in the institutional processes through which these allocation decisions are taken. The drive of scientists in pursuing their professional interests, the way in which educators and government bureaucrats perceive their responsibilities, and their success in persuading the community at large to share their vision — these are likely to be the dominant factors in determining the rate of expansion and diffusion of knowledge.[3]

This is not to say that traditional economics has nothing to contribute under these circumstances. Work done in recent years on the economics of education, or "investment in human capital," demonstrates that economic analysis can yield quantitative measures of returns to investment in educa-

---

[2]This point is stressed by Richard R. Nelson, "The Simple Economics of Basic Scientific Research," *Journal of Political Economy*, June 1959. See also Schultz, *Transforming Traditional Agriculture*, p. 150.

[3]"It may be true that in these days the search for new ideas and techniques is pursued with more system, greater energy, and, although this is more doubtful, greater economy. Yet chance still remains an important factor in invention, and the intuition, will and obstinacy of individuals spurred on by the desire for knowledge, renown or personal gain the great driving force in technical progress." J. Jewkes, D. Sawers, and R. Stillerman, *The Sources of Invention* (London: The Macmillan Co., 1958), pp. 223-4, quoted in S. Enke, *Economics for Development* (Englewood Cliffs, N.J.: Prentice-Hall, Inc., 1963), p. 100.

tion, which should enrich public discussion of the issues involved in this field.[4] The development of "cost effectiveness" techniques and their application to the analysis of defense and other governmental activities show that economic reasoning can be useful in areas perhaps even more remote than education from the marketplace.

Clearly, the analysis of costs and benefits in these areas is feasible and useful. This is not enough to explain the pattern of resource allocation, however, unless it can be assumed that resources will flow into those uses where rates of return are highest. This assumption is clearly implicit in the statement of Professor Schultz that "Once having determined the underlying costs and the return to each activity [i.e., experience, on-the-job training, and schooling as alternative ways of acquiring new knowledge and skills], the rate of return on the investment that each entails can be estimated. *The differences in the rate of return would then be the indicators to guide private and public investment decisions in this area.*"[5]

The crux of the matter lies in the second sentence. Public investment decisions are made by people whose personal fortunes do not hang directly on the outcome of those decisions. These people are therefore under no financial compulsion to opt for the alternative with the highest rate of return. They are free to disagree with the market's judgment of the worth of education (and other activities) and to allocate resources according to their own value criteria. In the context from which the above quotation was taken, Professor Schultz asserts that there has been underinvestment in the education of farm children in the United States and that a major reason for this is lack of relevant economic information concerning the returns to such investments. The assigned reason may be correct. One would expect that if the rates of return to education were better known this would influence investment decisions in this area. But so long as the market penalties for "wrong" allocation decisions do not apply, there is no assurance that market criteria will dominate these decisions.

The growth of basic scientific knowledge and of man's capacity to apply it directly in production processes are perhaps the most obvious examples of important productivity-increasing inputs, the behavior of which is not readily explicable with the received analytical apparatus of economics. Given the expansion rate of knowledge and skills, however, the rate of flow of newly

[4]There now is a considerable literature on the economics of education. Notable contributions are T. W. Schultz, "Education and Economic Growth," in *Social Forces Influencing American Education* (Chicago: National Society for the Study of Education, 1961); Mary Jane Bowman, "Human Capital: Concepts and Measures," in *Money, Growth and Methodology: Festkript for Johan Akerman* (Lund, Sweden: Gleerup, 1961); Gary Becker, "Investment in Human Capital," *Journal of Political Economy*, LXX (1962), Part 2. This entire volume of the *Journal* is devoted to the economics of education. It includes articles by Schultz, Jacob Mincer, and others. See also Becker's *Human Capital* (New York: Columbia University Press, 1964).

[5]Schultz, *Transforming Traditional Agriculture*, p. 173. Emphasis added.

produced inputs into agriculture – or any other sector – should generally be more amenable to traditional analysis. It may well be found that the rate of adoption of a new fertilizer, an improved seed variety, or a better plow is adequately explained by reference to the cost-return ratios of these inputs in relation to those they threaten to replace.[6]

Even in these cases, however, other factors can be important in shaping allocation decisions. Policy makers may decide that the country's best development strategy calls for a greater flow of resources into industry and a lesser flow into agriculture than would occur in response to relative prices of inputs and outputs. The subsequent expansion of agricultural production and productivity would therefore be less than would have been predicted from the analysis of market forces alone.

Nonprice rationing of resources combined with systems of price controls is a likely possibility, and, where it occurs, price analysis alone will obviously not provide an adequate explanation of resource allocation. In the developing countries many of the productivity-increasing inputs for agriculture – machinery, fertilizers, and pesticides, for example – must be imported. In most of these countries the shortage of foreign exchange is such that all imports are closely controlled through a variety of techniques, including rationing of foreign exchange. In this case, explanation of the rate of incorporation of new inputs into agriculture requires analysis of import-control policies and their implementation as well as study of relative prices.

Inputs are rationed also through the use of credit policies. Most countries in the world today, whether developed or "developing," consider management of the financial system a crucial aspect of development policy. The key element in financial management, of course, is the amount of monetary, or credit, expansion. In the case of credit, this amount could be controlled by permitting interest rates to find the level dictated by supply and demand. In countries where inflation is endemic, however, the "equilibrium" interest rate is likely to be at a politically unacceptable level, since inflation encourages indebtedness and discourages saving. Instead of permitting interest rates to determine the amount of credit, therefore, governments ration it through administrative processes. The result is that the demand for credit, and farm inputs financed with it, is greater than supply at existing interest rates and input prices. Analysis of input markets will reveal that marginal returns exceed prices. Periods of temporary disequilibrium aside, and in the absence of monopsony power in input markets, traditional economics cannot account for this phenomenon. A complete explanation requires analysis of govern-

---

[6]Griliches provides a convincing demonstration of this in his study of hybrid corn in the United States, and he suggests that the proposition may also apply to the adoption of cornpickers, tractors, and fertilizers. Zvi Griliches, "Hybrid Corn: An Exploration in the Economics of Technological Change," *Econometrica*, 25 (1957), 521–22.

ment financial policies and of the financial institutions through which they are implemented.

Traditional theory also encounters trouble wherever social attitudes yield different marginal valuations of resources. This is particularly important in agriculture, where the mystique of land ownership may persuade farmers to accept marginal returns to land substantially below those obtainable from other resources possessing less "magic." Farmers may also accept relatively low returns to land for wholly mundane reasons, fully consistent with economically rational behavior. Land ownership may be the key which permits access to agricultural credit. In this case, the relatively low return to land could be wholly or more than compensated by the high returns to nonland inputs not obtainable without credit. Although farmers' behavior is economically rational in these circumstances, it is not explicable through price analysis alone. Study of the relationship between land ownership and credit availability is essential.

The discussion up to this point yields three principal conclusions:

1. The rate of increase of agricultural productivity depends upon the rate at which improved inputs are incorporated into the production process at the farm level. These inputs include human skills as well as produced goods.

2. The study of agricultural productivity, therefore, requires analysis of the mechanisms which control the rate of invention and adoption of new inputs.

3. These mechanisms include those traditionally studied in economics — markets in which prices of inputs and outputs are set. To this extent, ordinary economic analysis is useful in studying the process of agricultural productivity changes. However, nonprice factors will frequently be important in governing the rate of incorporation of productivity-increasing inputs. In these cases traditional economic analysis must be supplemented by study of the mechanisms through which these nonprice factors exert their influence.

These conclusions have important methodological implications for research concerned with agricultural development and productivity. They indicate that the scope of the analysis must be broad enough to include systematic investigation of the role of nonprice, institutional forces in shaping resource allocation decisions. This in no way asserts that prices are not important. They undoubtedly are, and in the study of Chilean agricultural development, the traditional tools of economic analysis proved indispensable. It will become clear in the following chapters, however, that nonprice forces also exerted an important influence on the pattern of resource allocation in the agricultural sector. In particular, the conditions of land tenure, rural education, and government policies with respect to credit and foreign trade had a major impact on the ability and incentives of farmers to expand the employment of such high-productivity inputs as fertilizers, pesticides, and farm machinery. The analytical framework adopted, therefore, while relying heavily on price analysis, also provides for systematic consideration of the effects of nonprice institutional factors in shaping the investment and production decisions of Chilean farmers.

Institutional Framework
of Input Supply

The pace of agricultural development depends upon a host of institutional and economic conditions affecting the ability and willingness of farmers to adopt new inputs, thus expanding production and increasing the productivity of land and labor. In subsequent chapters the growth in the 1950's of production and of employment of various inputs, both in the country as a whole and among a sample of farms in the Central Valley, is examined in some detail. Those chapters attempt to establish quantitative indicators of agricultural performance and to explain it. Before taking up that material, however, it will be useful to examine the general characteristics of input demand and supply conditions that prevailed in the decade of the 1950's. This preliminary sketch will facilitate considerably the presentation and analysis of the quantitative material discussed in Chapters 3, 4, and 5.

GENERAL CONSIDERATIONS

The kinds and quantities of inputs employed by farmers reflect forces originating on the demand side of the market as well as conditions of input supply. On the demand side the important factors are the rate of growth in demand for farm products and the prices at which these can be sold in relation to the prices of the inputs required to produce them. On the supply side attention must be focused on the productivity, or quality, of inputs and on the conditions that determine the amounts of them which will be made available at various prices.

The demand for farm commodities produced in Chile grew considerably faster than production in the 1950's, the widening gap between demand and supply being filled by imports. This indicates that Chilean farmers could have sold considerably more than they did at prevailing product prices had the necessary inputs been available at favorable prices. Hence the limits to the expansion of output and to the increase in quantities of inputs employed were set by the price and other conditions of input supply. Apparently Chilean farmers did not expand production in the 1950's in pace with rising demand because these price and other supply conditions (including the farmers' own attitudes toward work and leisure) did not favor the necessary rise in the employment of inputs. In the discussion that follows, therefore, attention will be focused on the price and other supply conditions of inputs in the 1950's. Separate consideration is given to each of the major inputs. However, there were a number of factors whose influence was felt generally throughout most input markets, namely the price, credit, and foreign-trade policies pursued by the Chilean government in the 1950's. It is convenient to begin the discussion, therefore, with consideration of these policies.

Price Policies

Until the 1930's prices of agricultural products and inputs in Chile were set by market — predominantly world market — forces. The disruption in world and national markets caused by the Depression led to the beginning of governmental efforts to regulate prices. These became more vigorous and widespread with the coming of World War II, and by 1950 price controls of various sorts were applied to practically all agricultural products and to most inputs.[1] This system continued until about 1955, when controls were relaxed somewhat, leaving prices of some products "relatively free."[2]

The policies employed to regulate prices were a mixture applied to both imported and domestically produced products and inputs. A multiple exchange system was in force up to the mid-1950's, and this kept prices of both imported foods[3] and some farm inputs (notably machinery)[4] below the levels which would have prevailed with a freely fluctuating exchange. Domestically produced commodities were subject to official price controls, special taxes and subsidies, and rationing, as well as official buying and selling in selected cases.[5] These controls had various objectives, but among the most important was to keep down the prices of food, which weighed very heavily in the

---

[1] Kurt Ullrich B., *Algúnos aspectos del control de comercio de la agricultura chilena, 1950–1958* (Santiago: Ministerio de Agricultura, 1964), p. 13.

[2] Ibid.

[3] Instituto Agrario de Estudios Económicos (INTAGRO), *Trabajos realizados entre los años 1960 a 1962*, II (Santiago, 1963), 378–86.

[4] Naciones Unidas, Consejo Económico y Social, *El uso de maquinaria agrícola en Chile: Estudio preparado por la División Agrícola Conjunta CEPAL/FAO con la colaboración del Banco Interamericano de Desarrollo*, Doc. E/CN.12/799 (Santiago, Feb. 14, 1968), p. 27.

[5] Ullrich, *Algúnos aspectos*, p. 17.

cost-of-living index. Thus the usual practice was to set maximum rather than minimum prices for farm products.[6]

It is very difficult if not impossible to disentangle the effects of these various policies from all other forces influencing prices of agricultural products and inputs in the 1950's. The problem, difficult enough in any case, is compounded by the fact that various sources offer differing estimates of the behavior of prices over this period. Table 1 gives an example of this, indicating that by one set of estimates, wholesale prices of agricultural commodities increased 17.5 times between 1951 and 1960. The other estimate indicates an increase of only 15.1 times. One estimate shows the prices of tractors up 26.8 times over this period while the other reports a rise of 32.8 times.

Table 1. Relative Increases in Prices of Agricultural Commodities (at Wholesale) and Tractors

	Agricultural commodities		Tractors	
	(1)	(2)	(1)	(2)
1951–55	5.0	4.3	5.5	5.5
1955–60	3.5	3.5	4.9	6.0
1951–60	17.5	15.1	26.8	32.8

Sources: The columns (1) are estimates of the Corporación de Fomento (CORFO), the government agency responsible for promoting development, cited in *El uso de maquinaria*, p. 29. The columns (2) are from Ministerio de Agricultura, Departamento de Economía Agrícola, *La agricultura chilena en el quinquenio 1956-1960* (Santiago, 1963), Table 123 (facing p. 152) and p. 166. Actually, the index on p. 166 is for "machinery," but this includes only tractors and plows. The former account for about 93 per cent of their combined value.

These differing estimates indicate the need for caution in discussing changes in prices of agricultural commodities and inputs in the 1950's. They are not so divergent, however, as to make such discussion irrelevant. Accordingly, Table 2 presents estimates of relative price movements between 1951 and 1960. It appears that wholesale prices of agricultural products rose somewhat more rapidly than input prices between 1951 and 1955. This movement was reversed in the second half of the decade, and by 1960, input-output price relations were about the same as in 1951. Column (5) is an index representing all inputs except labor. It is very heavily weighted with fertilizers and machinery. Comparison of this index with those for product prices (columns (1), (2), and (3)) and for all inputs (column (4)) indicates

[6]María Inés Olivos C., Sergio Figuero T., and Jaime Roman V., *La estructura institucional: Un nuevo ángulo para analizar el problema del estancamiento de la agricultura chilena* (Santiago: Universidad de Chile, 1967), pp. 94–107. (Thesis presented for the degree in agricultural economics.)

that relative price movements must have tended to discourage the more rapid incorporation of "modern" inputs into Chilean agriculture in the 1950's. Not only did prices of these inputs rise much faster than product prices, they also rose more rapidly than the price of labor. This latter point is implicit in the movements of the indexes in columns (4) and (5).[7]

Table 2. Indexes of Prices of Agricultural Products and Inputs

(1956-60 = 100)

Year	Indexes of prices of agricultural products			Indexes of prices of agricultural inputs		Relative prices	
	All	Crops	Livestock	All	All except labor	Column (1) ÷ (4)	Column (1) ÷ (5)
	(1)	(2)	(3)	(4)	(5)	(6)	(7)
1951	9.5	9.8	9.1	9.2	6.8	1.03	1.40
1952	12.9	14.0	11.6	12.3	9.6	1.05	1.34
1953	15.1	15.4	14.7	17.2	13.1	0.88	1.15
1954	24.5	22.8	26.7	24.0	22.0	1.02	1.11
1955	41.2	36.6	47.4	35.9	29.7	1.15	1.39
1956	59.1	56.9	62.0	55.0	49.6	1.07	1.19
1957	81.0	84.8	75.8	81.7	86.2	0.99	0.94
1958	94.8	95.7	93.5	95.8	102.7	0.99	0.92
1959	122.1	116.8	129.3	126.8	123.9	0.96	0.99
1960	143.0	145.8	139.3	140.8	137.6	1.02	1.04

Sources:
    Columns (1) to (4) are from *La agricultura chilena... 1956-1960*, Table 123 (facing p. 152) and Table 130 (p. 166).
    Column (5) is constructed by the author from information contained in the same work, pp. 166-67.

Note: "All" agricultural inputs include seeds (wheat and potatoes), animal feed, nitrate and phosphate fertilizers, tractors and plows, petroleum (fuel), and labor. The latter is valued at the official minimum wage in Santiago Province. In the case of fertilizers, the price is net of subsidy.

Aside from movements in relative prices, a factor of major importance was the haphazard fashion in which price policies were set and administered. This point receives considerable emphasis in the agricultural plan for 1965-80. Throughout the 1950's responsibility for price policies was spread among various government agencies, making the development of consistent policies difficult if not impossible. Moreover, all too often prices were set in response to pressures from special interests, with little heed paid to the consequences for production. Because governments were reluctant to feed speculation

[7]A word of caution here. The "price" of labor is represented by the official minimum wage in Santiago Province. Actual wages may have risen faster, or more slowly, than the official wage. They would have had to rise very much faster, however, to alter the pattern shown in Table 2.

about the likely pace of inflation or, perhaps, to indicate acceptance of continued inflation, prices generally were set late in the production year. Consequently, in making their production decisions farmers were forced to make their own guesses concerning future prices. Of course, farmers everywhere usually have to do this. In the Chilean case, however, the situation was especially difficult because political factors frequently outweighed economic ones in price setting.[8]

Credit

Chilean credit policy in the 1950's exhibited a rather fundamental inconsistency. The country entered the decade still suffering from the chronic inflation which had plagued it for decades. As a move toward controlling this the Central Bank in mid-1953 was given authority to impose quotas on the rate of expansion of bank credits, to set bank reserve requirements, and to modify the terms on which it would rediscount bank paper. The Bank's initial use of these powers was hesitant, if not timid, but by 1955 the inflation showed signs of "running away" (the consumer price index that year rose by 76 per cent after an increase of 56 per cent in 1954) and the country at last seemed willing to move more strongly to contain it. The economic consulting firm of Klein and Sachs was invited to send a mission to advise on measures to do this. Among other things the mission recommended that the Central Bank move much more aggressively to limit the expansion of credit, which it did. This and other measures proved effective in slowing, although they did not stop, the rate of inflation, and in 1960 consumer prices rose only about 12 per cent.

The Central Bank's efforts to control the expansion of credit were frustrated to a considerable extent because throughout the decade it was called upon to finance large deficits in the Federal budget, thus inevitably increasing the lending capacity of the banking system. However, the Bank's efforts, plus the increasing share of total credits taken by the public sector, did result in a sharp drop in the *real* value of credits extended to the private sector. Deflating the current value of private sector credits by the index of the general price level indicates that the real value of these credits, expressed in prices of 1960, fell by 47 per cent from 1951 to 1959.[9]

Loans to agriculture shared fully in the decline in the real value of private sector credits, as Table 3 shows. The sources of these loans were private and public banks and other lending institutions, such as the Corporación de Fomento (CORFO) and the Chilean land reform agency. Between them, these

[8]This paragraph is based on an unpublished memorandum made available by the Oficina de Planificación Agrícola (ODEPA).

[9]Ministerio de Agricultura, Departamento de Economía Agrícola, *La agricultura chilena en el quinquenio 1956-1960* (Santiago, 1963), p. 243.

Table 3. Credits Extended to Agriculture and Other Sectors, 1951–60

Year	Credits extended (Million 1960 E°) Total	Agriculture	Percentage shares of total Agriculture	Industry	Commerce	Others[a]
1951	586.5	220.4	37.6	23.8	26.6	12.0
1952	529.8	175.4	33.1	28.9	27.3	11.3
1953	549.4	154.9	28.2	30.8	30.5	10.5
1954	405.0	128.9	31.8	34.4	25.3	8.5
1955	408.2	142.2	34.9	35.5	22.4	7.2
1956	326.6	127.9	39.2	33.7	20.9	6.2
1957	310.1	105.2	33.9	37.0	22.9	6.2
1958	305.4	111.2	36.4	35.3	22.0	6.3
1959	290.6	117.8	40.5	30.8	21.7	7.0
1960	471.0	162.0	34.4	33.4	25.9	6.3

Source: *La agricultura chilena... 1956–1960*, p. 243. This source deflates loans in escudos (E°) of each year by the Index of the General Price Level to estimate the real value of loans. We accepted this procedure, except that for agriculture we deflated by the price index of nonlabor inputs shown in Table 2, column (5). Thus total loans in this table differ somewhat from those in *La agricultura chilena... 1956–1960*. The difference is quite small, however. In 1960 the official exchange rate for the escudo was E°1 = $.95.

[a]Includes private individuals, professions, and real estate and other brokers.

bodies accounted for all so-called institutional lending. Excluded are the transactions between persons.[10]

It is frequently asserted that the credit system in Chile discriminates against agriculture in favor of other sectors, particularly industry. This appears not to have been the case in the 1950's, however. Agriculture's share of total loans made in this period fluctuated somewhat, but there was no persistent tendency to lose ground. Indeed, agriculture captured a somewhat larger proportion of total loans in the second half of the decade than in the first. It is true that industry gained, but at the expense of "other" borrowers and the commercial sector rather than agriculture.[11]

However, the real value of the loans granted to agriculture fell drastically and persistently over most of the decade, reaching a level in 1959 only half of that of 1951. As a consequence, agricultural loans fell sharply as a percentage of the value of agricultural production in the 1950's. This decline in the real

[10]Lending between individuals is important in agriculture, and while no nationwide data are available on these transactions, some useful information is contained in a study by Nisbet. This is treated below.

[11]It is commonly believed that some loans which appear in the statistics as "agricultural" are in fact channeled into other sectors. This may well be true; however, it would not affect the conclusion that agriculture maintained its relative position, unless the percentage of such diversions increased over the decade. There is no apparent reason to believe this happened.

Table 4. Distribution of Agricultural Loans by Source in the 1950's

						(Percentage)
			Average			Average
	1951	1952	1951–55	1959	1960	1956–60
Banks:						
Private	30.8	30.9	32.4	30.7	34.2	30.8
Public:						
Central Bank	3.3	3.4	2.8	5.9	4.6	5.4
Banking Dept., State						
Bank	20.5	22.7	21.5	17.8	16.0	18.0
Development institutions:						
Agricultural Dept.,						
State Bank	25.7	24.5	31.0	39.4	34.3	37.7
Land Reform Agency	3.0	2.2	1.6	1.6	2.5	2.0
CORFO	0.8	1.1	1.3	2.0	5.7	3.3
Mortgage institutions:						
Mortgage Dept., State						
Bank	9.4	9.2	3.9	0.5	0.4	0.4
Mortgage Bank of Chile	5.0	4.7	3.6	1.7	1.9	1.9
Mortgage Bank of						
Valparaíso	1.4	1.3	1.1	0.5	0.4	0.5

Source: La agricultura chilena . . . 1956-1960, p. 245.

value of agricultural loans unquestionably slowed down the rate of incorpora-tion of machinery, fertilizers, pesticides, and other inputs purchased with credit.[12]

The distribution of agricultural credits according to source is given in Table 4. Private commercial banks were the single most important source in the first half of the 1950's, and throughout the decade they provided about 30 per cent of the total of such credits. Moreover loans to agriculture ac-counted for about 21 per cent of all loans extended by private banks in that period.[13]

Private-bank lending to agriculture was very short term, almost all loans being for less than a year. Many of them were designed to finance harvesting and selling of products.[14] Nominal interest rates charged ran between 15 and 20 per cent annually in the 1950's and averaged between 17 and 18 per cent.[15] Standard banking criteria were applied in judging the credit-worthi-

[12] The relation between credit availability and the employment of these inputs receives more attention in the following sections.

[13] *La agricultura chilena . . . 1956-1960*, p. 246.

[14] Ibid.

[15] INTAGRO, *Trabajos realizados,* I, 4. These interest rates are called nominal be-cause they were well under the annual increase of the price level in the 1950's. Hence

ness of loan applications, meaning that the stronger the asset position of the applicant the better his chances of obtaining credit. Larger farmers, therefore, were in a stronger position than smaller farmers in the competition for private-bank credit.

The Banking Department of the State Bank, which accounted for about 20 per cent of total institutional credits extended to agriculture in the 1950's, lent on the same basis as the commercial banks.[16] Thus about 50 per cent of all institutional loans to agriculture in the 1950's were very short term (less than one year) and were placed according to criteria which distinctly favored larger farmers over the smaller ones.

The Agricultural Department of the State Bank was the principal source of so-called "development" credits to agriculture in the 1950's. These are loans that may run up to 3 years. In the period 1956–60, 36.5 per cent of the loans of the Agricultural Department were to finance fertilizers and pesticides; 25.4 per cent, annual crops (15.6 per cent, wheat); 17.5 per cent, permanent on-farm improvements; 10.4 per cent, purchases of machinery; and 7.7 per cent, livestock operations; while plantations and vineyards got only about 1 per cent.[17]

The Department granted loans both in cash – to finance purchases in the open market – and in kind, i.e., from its own stocks of farm inputs. The overall amount of cash loans that could be made was determined by the Superintendent of Banks and by the Central Bank, acting within credit policy guidelines set for the country as a whole.[18] The amount of lending in kind was set by the amount of merchandise which the Department was authorized to purchase for resale on credit, by the amounts of specific kinds of goods available, by the capacity of the Bank's warehouses, and by the debt-carrying capacity of farmers.

Only individuals and organizations showing evidence of land ownership or land rental contracts were entitled to credit. The Bank distinguished between large and small farmers, the latter being those whose net capital did not exceed E°4,000 (about US$3,800 in 1959). Loan applications by small farm-

---

real rates of interest were negative. The private banks nevertheless were able to live with this situation for two reasons: (1) loan fees discounts and compensating balances offset some of the difference between nominal and real rates; and (2) Central Bank financing of Federal government deficits steadily expanded the lending capacity of the banking system, thus permitting total bank income to rise more or less in step with rising costs, despite the negative interest rates.

[16] *La agricultura chilena . . . 1956–1960*, p. 247.

[17] Ibid., p. 251.

[18] This and subsequent statements concerning the operations of the Agricultural Department of the State Bank are based on Ernest Feder, *Controlled Credit and Agricultural Development in Chile* (University of Nebraska, Aug. 1959, mimeo.). This study is the result of a six month investigation of the Agricultural Department's operations carried out in the first half of 1959.

ers, tenants, and sharecroppers were supposed to be judged by more than strictly commercial criteria. These included the farmer's past performance in debt payment, his farming ability, and his reputation for sobriety and perseverance. In fact, however, the Bank showed "little evidence of carrying out this professed policy."[19]

Bank loans, whether in cash or in kind, typically carried interest rates of 10 to 15 per cent, well below the prevailing commercial-bank rates and even further under the annual rate of inflation. The Bank preferred lending out of its own stocks of commodities. Repayment periods for these loans varied all the way from 30 days after harvest up to 2 years, depending upon the purpose of the loan.

Cash loans were supervised rather strictly to assure that they were used for the stated purpose and that the pledged collateral was legitimate. Generally, these controls appeared to be more strictly applied to small farmers than to large ones.[20]

Lending decisions were highly centralized, some 60 to 70 per cent of the value of all loans being granted from Bank headquarters in Santiago. Some 60 to 75 per cent of the value of loan requests were actually granted. Generally speaking, large loan requests were scaled down proportionately more than smaller ones. Thus, of the loan applications considered by the Bank's Executive Committee, 28 per cent of those exceeding $E^o2,000$ were scaled down 50 per cent or more while only 6 per cent of applications for less than $E^o2,000$ were so treated.[21] Similarly, the loans granted by the Agricultural Committee were substantially smaller, on average, than those it rejected.

The Bank was particularly interested in the net capital position of prospective borrowers and in their outstanding debts to the Bank. These factors largely determined eligibility for a loan, its amount, rate of interest, and term of repayment, and the maximum amount of outstanding indebtedness permitted.

From among loan applications considered by the Executive Committee, Feder chose a random sample of 163 submitted by 137 applicants and examined them in detail. He reports that a survey of the names included in the sample revealed a large number of well-known personalities in industry, commerce, the professions, and politics, and observes that "considering the small size and randomness of the sample, it is quite remarkable that part of the list of the Bank customers should read like a social register."[22] A high proportion of the customers requesting loans controlled substantial holdings of land and other resources, and in general the value of these holdings was grossly under-

[19] Ibid., p. 88.
[20] Ibid., pp. 18–19.
[21] Ibid., pp. 29–30.
[22] Ibid., p. 38.

stated in the loan applications.[23] Feder concludes that "it may, therefore, be hazarded to say that many, if not the majority, of the potential recipients of loans from the Bank are the 'upper crust' of Chilean farmers."[24]

Only 43 of the 137 borrowers included in the sample had net capital of less than $E^o 25,000$. Feder interprets this to mean that the Executive Committee generally dealt more harshly with smaller loans. Hence, smaller farmers simply limited their requests to amounts they knew were within the lending limits of Bank branches, thus avoiding confrontation with the Executive Committee. (How this squares with the earlier statement that large loan requests were scaled down proportionately more than small loans is not explained.)

In his conclusion, Feder estimates that the total number of Bank customers for agricultural loans was perhaps 30,000 farmers. "But of these, more than half probably obtain no more than very small credits."[25] If these estimates are correct, it is likely that as much as one-third of the total value of Bank loans to agriculture were granted to no more than 1 to 2 per cent of the farmers in the country.[26]

The only other two institutions extending nonmortgage credits in the 1950's were the agrarian reform agency and CORFO. Quantitatively, neither was of much significance, although CORFO played an important role in financing machinery imports. This is dealt with below.

Mortgage loans ceased to be a factor in agricultural credits in the 1950's. Whereas at the beginning of the decade almost 16 per cent of the value of loans granted were for mortgages, by 1960 this share had dropped to barely more than 2.5 per cent.

A study undertaken in 1965 has shown that there was a very large (in number of participants) and active market for so-called noninstitutional agricultural credits in Chile.[27] The borrowers were small farmers, most of whom would not qualify for institutional credits, and the lenders were mostly local merchants, professional money lenders, large landowners for whom the borrowers worked, friends, and relatives. Nisbet, whose study was based on a field survey of 200 farmers located in the Central Valley, found that in that area about 50 per cent of the farmers surveyed obtained noninstitutional credits. Because the Central Valley area is the agricultural heartland of the country, Nisbet reasons that it is better served by private and government banks than other areas. He concludes, therefore, that in the country as a

---

[23] Ibid., pp. 39, 42–45.

[24] Ibid., p. 48.

[25] Ibid., p. 84.

[26] Ibid., p. 86.

[27] Charles T. Nisbet, "Oferta de fondos financieros fuera del mercado crediticio institucional en el sector rural de Chile," *Revista de economía*, XXIV (1966), 3d quarter, pp. 29–42.

whole more than half of all farmers are in the market for noninstitutional credits.[28]

Loans were given in money or in kind, the latter being the more common. The range of commodities in this category covered practically every conceivable input as well as consumption goods.[29] The market area covered by any one lender fluctuated within a radius of 1 to 8 kilometers. On the average there were two lenders within each market area. Nisbet asserts that lenders had almost perfect knowledge of the financial condition of borrowers, but that the latter knew little or nothing about alternative lenders. Demand for credit was very inelastic with respect to interest rates since it was determined by the farmer's need to keep his farm in operation and to sustain himself and his family until the first harvest.[30]

Almost all the loans made by employers, friends, and relatives — 50 of 105 transactions — were nominally at zero interest rates. Nisbet argues that employers were willing to engage in this apparently uneconomic practice because it tied the workers more tightly to them, thus ensuring a steady supply of low-cost labor.[31] However, since it is the loan itself, not the rate of interest, which gives the employer a claim on the borrower's time, this argument is not very convincing.

Nominal interest charged by merchants and professional money lenders was zero in 2 of 55 transactions. In all others it ranged from 18 per cent to 360 per cent (annual rates). The modal interest charged was 60 per cent.[32] The term of the loans fluctuated between 6 and 8 months.

Real interest rates on loans made in cash were of course less than nominal rates because of inflation. In the extreme cases of cash loans given at zero nominal interest, the real rates were negative at 33 per cent. Real rates on loans given in kind were about the same as nominal rates, with 60 per cent the modal value.[33]

The noninstitutional credit market obviously consisted of two distinct parts. In one, loans were granted for reasons of friendship, family ties, or the prospect of indirect gain sufficient to offset losses from negative interest rates. The other was motivated by strictly pecuniary considerations, and interest rates probably contained an element of monopolistic or oligopolistic gain. Given the failure of regular credit institutions to reach the great bulk of small farmers, the noninstitutional sources no doubt played an important role in financing Chilean agriculture. The social cost of these credits was probably high, however, owing to the very small scale on which the typical lender

[28] Ibid., p. 32.
[29] Ibid., p. 33.
[30] Ibid., p. 34.
[31] Ibid., p. 35.
[32] Ibid., p. 36.
[33] Ibid., pp. 39–40.

operated. The cost to the individual borrower was even higher, as lenders were in a position to extract monopolistic or oligopolistic gains.

In summary, the real value of agricultural loans granted in the 1950's declined very sharply, although agriculture managed to maintain its position in relation to other sectors in the competition for credit. About one-half the loans granted were very short term, averaging under 1 year. Most of the rest were for terms of 1 to 3 years. The strength of the applicant's asset position was the principal criterion by which loans were granted, a feature which inevitably gave advantage to large farmers over small farmers in the competition for such modern inputs as fertilizers, pesticides, and machinery. Moreover, it permitted larger farmers more flexibility in planning production and marketing strategies. Nominal interest rates on these loans were under the prevailing rate of inflation.

The only significant institution making loans of more than 1 year was the Agricultural Department of the State Bank. Few of its loans, however, were for more than 2 years. Lending operations were highly centralized in Santiago, and in general large farmers, many of them socially and politically prominent persons with extensive interests outside agriculture, received more favorable treatment than small farmers. No more than 1 to 2 per cent of all the farmers in Chile received perhaps as much as one-third of all loans extended by the Agricultural Department. Small farmers who did not qualify for these credits were forced to do without or else to deal with private money lenders. Apparently many of them did the latter. Those lucky enough to have friends, relatives, or employers willing and able to lend to them could expect to repay less than the real value of the loan, owing to the failure of interest rates to compensate for the rate of inflation. However, those who had recourse to commercial noninstitutional lenders paid very high real interest.

In short, long-term lending to agriculture virtually did not exist in Chile in the 1950's, and even medium-term lending was on a relatively small scale. There was a pronounced bias in favor of the larger over the smaller farmer. The first feature tended to discourage longer-term investments in land improvements and machinery, while the second greatly strengthened the position of the larger farmer in the competition for modern inputs.

Foreign Trade

Throughout the 1950's the Chilean government was confronted with a balance of payments problem, its severity increasing or decreasing largely in accordance with movements in the world price of copper. As a consequence of this problem, imports were subject to control throughout the decade. Until early 1956, foreign-exchange quotas were established each year for various classes of imports, the amounts of the quotas depending upon official estimates of the total amount of foreign exchange likely to be available and judgments about the priorities of the various classes of imports. Persons or firms wishing to import were required to petition the Board of Trade (CON-DECOR) for permission to do so. In those cases in which permission was granted, CONDECOR also set the rate of exchange applicable to the transaction.

Early in 1956 the policy of foreign-exchange quotas was changed to one of indirect controls. Lists of permitted and prohibited products were drawn up. To import from the permitted list it was necessary to make a declaration of intent to the Foreign Exchange Commission and to deposit in the Central Bank a percentage of the value of the proposed imports. These percentages varied with the class of commodity, and the whole structure of percentages was varied from time to time depending upon the government's assessment of the need for more or less control. In 1957, for example, there was a sharp drop in copper prices and the balance of payments deteriorated badly. In an effort to slow the expansion of imports, the required deposit percentages were raised several times over the course of the year. So strong was the demand for imports, however, that the control measures did not really take hold until early 1958.[34]

In general, low or zero deposits were required for food and other agricultural commodities. The deposits required for agricultural inputs, however, were relatively high, fluctuating between 50 per cent and 150 per cent and on occasion going even higher.[35] While the quantitative impact of these controls on imports of agricultural inputs cannot be precisely determined, some evidence on this point is considered in Chapter 4.

Machinery[36]

It is shown in the following chapter that the stock of farm machinery increased by some 80 to 120 per cent in the 1950's, and that the return to this additional investment was on the order of 10 to 15 per cent. There is little question, however, that the real return was substantially higher than this. The reason is that much of this machinery was obtained by means of CORFO loans given at 10 per cent interest with no adjustment of outstanding indebtedness to allow for inflation. It is estimated that in the period 1953–56 farmers who managed to get these loans in fact paid back only 40 per cent in real terms. In addition the multiple-exchange-rate system in effect up until the mid-1950's was very favorable for farm machinery imports.[37]

The average real rates of return on private investments in farm machinery, therefore, must have been quite attractive in the 1950's. However, the dom-

[34] Banco Central de Chile, *Trigésima segunda memoria anual, año 1957* (Santiago, 1958), pp. 63–64.

[35] Ibid.

[36] This part draws very heavily on *El uso de maquinaria.*

[37] Ibid., p. 61.

inant factors controlling the rate of acquisition of such machinery were government credit and foreign-trade policy. The effects of relative prices of machinery were apparently swamped by the two nonprice factors. These statements are based on analysis of the relationship between imports of farm machinery on the one hand and prices and credit on the other. Some 90 to 95 per cent of the farm machinery employed in Chile is imported.[38] The value of such imports between 1950 and 1965 is indicated in Table 5.

Table 5.   Imports of and Credits for the Purchase of Agricultural Machinery in Chile, 1950–65

Year	Imports (thousand U.S. $)	Credits extended (thousand U.S. $)
1950	2,491	1,553
1951	7,969	1,893
1952	7,197	3,058
1953	9,776	2,772
1954	14,028	6,946
1955	17,020	4,750
1956	9,366	2,600
1957	11,925	3,696
1958	7,113	3,245
1959	5,519	2,137
1960	11,219	3,220
1961	15,609	7,152
1962	13,196	5,219
1963	12,722	4,860
1964	9,536	8,202
1965	11,674	6,232

Source: Machinery imports for 1953–65 are from El uso de maquinaria, p. 10. Imports for 1950–52 and credits for all years are from the files of the author of El uso de maquinaria.

Analysis of the relationship between these imports and their prices would require an index representing the movement in average machinery prices as well as an index of the quantity of machinery imports. Neither index was available. There is information, however, on the prices and numbers of tractors imported each year. Since tractor imports accounted for about 50 per cent of total farm machinery imports between 1950 and 1965, tractors are a reasonably good proxy for all farm machinery for purposes of this analysis.

The analysis showed that in the years 1950 to 1965 there was at most only a very weak correlation between tractor prices and the number of tractors imported.[39] This, of course, does not mean that farmers were insensitive to price changes, other things being equal. But "other things," specifically government credit and foreign-trade policies, were by no means equal in this

[38] Ibid., p. 27.

[39] The regression of number of tractor imports on the price of tractors was

$$Y = 2265.4 - 8.055X$$

period. Analysis of the relation between credits extended for the purchase of farm machinery and imports of these in the years 1950–65 revealed that credit explained about 44 per cent of the year-to-year variations in imports. However, the relationship was distorted in 1955 and 1964. When those two years are excluded credit explains about 74 per cent of the variations.[40]

The most likely explanation of the unusually large imports in 1955 is that it was generally believed that the system of multiple exchange rates, which favored the importation of farm machinery, would be changed, as indeed it was at the beginning of 1956. Besides this, it was generally known that by 1955 CORFO had suffered very severe capital losses through the extension of credits unadjusted for inflation, and farmers may well have judged that this policy would shortly be changed. In any case credits extended by CORFO and the State Bank rose to $6.9 million in 1954 from only $2.8 million in 1953. Then in 1955 CORFO's policy was changed and its loans were sharply cut back. From 1956 to 1959 it made no new loans, the State Bank being the only significant source of credit for imported farm machinery in this period.[41] Since there is a lag between the extension of credits and the importation of machinery, the build-up of credits in 1954 probably helps to explain the jump in imports in 1955.

---

where Y = annual number of tractors imported

X = index of the escudo price of tractors deflated by an index of the price of wheat

The data were for the years 1950–65 inclusive. The correlation coefficient was .298, significantly different from zero at about the 27 per cent level of probability.

A regression was done also of annual changes in the number of tractor imports on changes in price. This equation was:

$$Y = 117.9 - 19.1X$$

The correlation coefficient was .38, significantly different from zero at about the 17 per cent level of probability. Note that in both regressions the sign of the relationship was "right" despite the weakness of the correlation.

[40]The regression equation covering the years 1950-1965 inclusive was:

$$Y = 5.21 + 1.23X$$

where Y = c.i.f. value of annual imports of farm machinery measured in millions of dollars

X = the corresponding dollar values, in millions, of all credits extended by CORFO and the State Bank for purchase of farm machinery imports.

The correlation coefficient was .66, significantly different from zero at less than the 1 per cent level of probability.

When 1955 and 1964 were omitted from the regression the equation became:

$$Y = 3.46 + 1.67X$$

The correlation coefficient was .86, significantly different from zero at less than the 1 per cent level of probability.

[41]Lending by private banks and importers or distributors of machinery was on a very small scale and was always very short term. *El uso de maquinaria*, p. 59.

There is no obvious explanation for the failure of imports and credit to move together in 1964. The decline in imports may reflect the impact on farmers' investment decisions of uncertainty concerning the election of that year. This impact may have been sufficiently strong to more than offset the considerable expansion in credits, all of it made available by CORFO.[42] An additional factor may be that the increase in CORFO credits was so sharp that imports simply could not keep up. This view is given some credence by the dissimilar movement of imports and credits in 1965. The increase in imports in the face of declining credits may in part reflect the large rise in credits the previous year.

Thus government credit and foreign-trade policies appear to have been major factors in determining the pace of farm mechanization, almost completely eclipsing the effect of machinery prices. It will be shown in Chapter 4 that this is a conclusion of considerable importance in evaluating the performance of Chilean agriculture in the 1950's.

As noted above, CORFO and the State Bank were the only significant sources of farm machinery credit in the 1950's. Over the decade as a whole, the State Bank granted about 71 per cent of total farm machinery loans, and CORFO was responsible for the remaining 29 per cent. However, in the first half of the decade, before it withdrew completely from this activity, CORFO's share was 48 per cent. CORFO loans were granted for from 3 to 5 years. As indicated earlier, interest was 10 per cent and the loan balance was not adjusted for inflation. The State Bank's credits were generally for only 2 years. While they were apparently not adjusted for inflation,[43] interest rates, commissions, and fees apparently were sufficiently high to maintain the Bank's capital intact.[44] Farmers were required to finance with their own capital 30 per cent of the sales price of machinery purchased.

The pronounced tendency of the State Bank to confine its lending to larger farmers has been noted previously. Farm machinery loans were no exception to this policy. CORFO also limited its credits to large and medium-sized farmers by insisting on a strong capital position as a prerequisite for obtaining a loan.[45]

Hence smaller farmers were effectively excluded from the market for new farm machinery. Little is known about the second-hand market, although there obviously was one, and smaller farmers had access to it. However, the machinery thus available was of course older, perhaps much older, and, by

[42] CORFO accounted for about two-thirds of all farm machinery credits extended in 1963. In 1964 CORFO more than doubled these credits (all in prices of 1966), raising its share of the total to 80 per cent. Then in 1965 the agency reduced its lending by one-third. Ibid., p. 66.

[43] *El uso de maquinaria* is not clear on this point.

[44] Ibid., p. 67.

[45] Ibid., p. 64.

virtue of this, less efficient than that being introduced on larger farms. The smaller farmers were thus disadvantaged in relation to the larger farmers.[46]

The smaller farmer's plight in this respect was alleviated somewhat by the activity of SEAM (Servicio de Equipos Agrícolas Mecanizados), an arm of CORFO. SEAM had three functions: to provide machinery services to small farmers at less than cost, to provide such services to medium-sized and large farmers at cost, and also to engage in the building of roads and small dams, in land clearing and leveling, and in other infrastructure tasks. In 1952/53 SEAM provided 134,000 hours of machine services, in 1955/56, 276,000 hours, and in 1958/59, 219,000 hours.[47] How much of this was provided to small farmers cannot be determined from the available data. However, in 1955 there were 116,000 farms at or below the size judged as adequate to support a family.[48] If all of SEAM's services had been rendered to these farms, they would have received an average of 2.4 machine hours in the course of the year. Since in fact these small farms must have received substantially less than 100 per cent of SEAM's services, the amount of machine time available to them was obviously quite small.

Efficient employment of farm machinery requires that spare parts be available without undue delay and at prices which do not absorb the profits to be had from mechanization. In addition, efficient repair services and people skilled in the operation of machinery are essential. Spare parts for farm machinery have always been expensive in Chile. The more important of them, such as motors, transmissions, and parts thereof, must be imported. In a detailed study of the operations of four importers of spare parts it was found that in the first half of the 1960's the list prices of parts ran 3.6 to 5.3 times the c.i.f. prices.[49] A comparison was made of the prices of certain imported motors and parts in Chile with those in other Latin American countries and the United Kingdom. In only one case — prices of crankshafts in Argentina — were Chilean prices less than those in the other countries.[50]

Since 1942 CORFO has had a joint program with the Chilean army for the training of tractor drivers and other operators of farm machinery. From that date until 1965, some 5,000 soldiers had passed through this program and subsequently returned to civilian life. Unfortunately, many of them took up

[46]Whether smaller farmers could have made effective use of more machinery is another question, treated in Ch. 5.

[47]*El uso de maquinaria*, p. 69.

[48]Comité Interamericano del Desarrollo Agrícolo (CIDA), *Chile: Tenencia de la tierra y desarrollo socio-económico del sector agrícolo* (Santiago, 1966), p. 304. (Hereinafter called CIDA Report.) The size of farm judged adequate to support a family varies with the type of soil and climate. The figure of 116,000 does not include the small plots within larger farms which are set aside for the use of laborers as part of the terms of their employment.

[49]*El uso de maquinaria*, p. 56.

[50]Ibid., p. 55.

nonagricultural occupations for the very good reason that these were better paid.[51] It was concluded that as late as the mid-1960's around 90 per cent of the machinery operators and mechanics on farms lacked the means and the knowledge necessary for efficient utilization of farm machinery.[52]

Fertilizers[53]

It is demonstrated in the following chapter that consumption of fertilizers in Chile about doubled over the course of the 1950's. Virtually all nitrate and about one-half of the potassic fertilizers consumed were locally produced.[54] In the early 1950's about two-thirds of the phosphates consumed also was produced domestically, but by the end of the decade this proportion had dropped to one-third, imports accounting for the balance. However, between 70 and 75 per cent of the *increase* in total fertilizer consumption over the decade was in imported phosphates and potassics, mostly the former.

The production and distribution of fertilizers consumed in Chile were powerfully influenced by governmental or quasi-governmental bodies in the 1950's. Nitrate production was largely in the hands of private firms, but in 1956 a government agency, COVENSA,[55] was given a monopoly in the purchase and distribution, both domestically and abroad, of all nitrate fertilizers. COVENSA set the price at which it would buy from the producers and also the price at which it would sell. The purchase price was based on the average production costs of all producers. The selling price was supposed to cover all distribution costs, and it could not be less than the average cost of production plus 10 per cent.[56]

The bulk of the marketing of imported phosphate fertilizers, which as noted above accounted for 70–75 per cent of the increase in fertilizer consumption in the 1950's, was in the hands of the State Bank. This organization was charged with soliciting bids from abroad and with distributing the product internally, both directly to farmers and through private firms.

The State Bank was also by far the most important source of credit for the purchase of fertilizers. Data for the 1950's are not available, but in the early 1960's fertilizer credits extended by the State Bank were about 65 per cent

[51] Ibid., p. 72.

[52] Ibid., p. 73. The figure of 90 per cent is a rule of thumb used by the technicians of CORFO, based on their experience in Chilean agriculture. It clearly is very rough and could be too high. The assertion that the great majority of farm machinery operators and mechanics is inadequately trained would be correct, however.

[53] This section is based heavily on Naciones Unidas, Consejo Económico y Social, *El uso de fertilizantes en Chile: Estudio preparado por la División Agrícola Conjunta CEPAL/FAO con la colaboración del Banco Interamericano de Desarrollo,* Doc. E/CN.12/757/Rev.1 (Santiago, June 17, 1966).

[54] Ibid., p. 4.

[55] Corporación de Ventas de Salitre y Yodo.

[56] *El uso de fertilizantes,* pp. 71–72.

of the total value of fertilizers consumed in those years.[57] As in its lending policies generally, the Bank's fertilizer loans tended to favor the larger over the smaller farmers.[58]

COVENSA granted loans for the purchase of nitrate fertilizers, making use of a line of credit extended to it by the Central Bank. These loans were typically for one year with interest at 12 per cent.[59] Information is not available on the size and other characteristics of the recipients of these loans.

Beginning in 1952 fertilizers received a subsidy equal to 50 per cent of the sales price after deduction of transport costs, taxes, and certain other costs. In subsequent years this subsidy was reduced, so that by 1956 it came to only 10 per cent. In 1957, 1958, and 1959 no subsidy was paid for lack of funds. In 1960, however, it was reestablished, reducing the effective prices of phosphates, nitrates, and potassics by about 45 per cent, 33 per cent, and 20 per cent, respectively. Consumption increased somewhat in 1960 but climbed much faster thereafter, so that by 1965 total consumption was fully 125 per cent above the 1959 level.[60] The real prices — nominal prices net of subsidy deflated by the wholesale price index of farm commodities — rose gradually over this period and by 1965 were from 10 to 25 per cent above their 1960 levels.

According to *El uso de fertilizantes*, the quantity and quality of technical studies of the fertilizer needs of Chilean soils, while leaving something to be desired, are nonetheless much better than would be thought at first glance. The main problem is that few of these studies have been published.[61] There are almost no studies of the economics of fertilizer use, a particularly serious failing since government policy with respect to subsidies was made with little if any knowledge of the amount needed to stimulate increased use of fertilizers, or of whether any subsidy at all was necessary. As a consequence, the subsidy has been determined by the pulling and hauling among government, distributors, and producers, all equally uninformed about the crucial issue.[62]

In the mid-1960's there were more than half a dozen public and private agencies providing technical information concerning fertilizer use to Chilean farmers. Yet their efforts were judged to have been inadequate, in part because of the quantity and quality of the resources available to them, but in part because of inadequate coordination among them, because too little had been done to relate research to extension, and because generally speaking the overall effort had been too passive. It had contented itself with simply making material available rather than aggressively bringing it to the farmers' atten-

[57] Ibid., p. 54.
[58] Ibid., p. 53.
[59] Ibid., p. 105.
[60] Ibid., pp. 5 and 44.
[61] Ibid., p. 31.
[62] Ibid.

tion and demonstrating its effectiveness in achieving higher productivity.[63] The agricultural extension effort in Chile in the mid-1960's was considerably larger and more vigorous than it was in the 1950's. Whatever its weaknesses in the 1960's, they undoubtedly were more pronounced in the previous decade.

Pesticides[64]

It is shown in the next chapter that consumption of pesticides in Chile rose between 3 and 4 times in the 1950's, although at the end of the decade the volume of consumption still was low. Most insecticides and herbicides were imported, but virtually all fungicides were domestically produced. Three firms accounted for about 96 per cent of total fungicide production and 94 per cent of all pesticides produced in the country.[65] In the early 1960's there were 65 firms importing a large number of different pesticides.[66] However, 5 of these firms accounted for 64 per cent of all imports, 7 for 75 per cent, and 18 for 94 per cent. Nevertheless, the number of firms and products was sufficiently large, and the rate of introduction of new products sufficiently rapid, to make the industry intensely competitive.[67]

A careful analysis of pesticide prices in Chile led to the conclusion that they were considerably higher than in developed countries (the United Kingdom was used as a standard of comparison).[68] This discrepancy was attributed to high internal distribution costs owing to the relatively small size of the market and the relatively large number of sellers and, especially, of products.[69]

Credit and technical assistance in the use of pesticides have been left almost wholly to private producers and importers. The State Bank was the only government lending institution active in this field, and in 1961 its loans amounted to only $E^{o}275,000$. The total value of pesticides consumed that year was estimated at $E^{o}3,637,000$.[70] Loans by private firms were extended for periods of 90 to 270 days, with interest ranging from 1.5 per cent to 2.0 per cent per month. These loans covered a "good part" of total pesticide consumption in the early 1960's.[71]

---

[63] Ibid., pp. 32–38.

[64] This section is based on Naciones Unidas, Consejo Económico y Social, *El uso de pesticidas en Chile: Estudio preparado por la División Agrícola Conjunta CEPAL/FAO con la colaboración del Banco Interamericano de Desarrollo,* Doc. E/CN.12/759 (Santiago, Sept. 5, 1966).

[65] Ibid., pp. 12–13.

[66] In 1963 about 450 pesticides were sold in Chile, although many of these differed only in brand name. Ibid., p. 6.

[67] Ibid., p. 2.

[68] Ibid., p. 43.

[69] Ibid., p. 50.

[70] Ibid., pp. 62 and 64.

[71] Ibid., p. 63.

Research and extension activities concerning the use of pesticides were virtually 100 per cent in the hands of private firms in the 1950's. These activities undoubtedly contributed to the substantial increase in pesticide consumption that occurred in that period.[72] The firms formed teams of technicians, who dealt directly with the farmers, instructing them concerning proper doses and application techniques. These services, however, were limited almost exclusively to large and middle-sized farmers, since either the smaller farmers lacked the resources to purchase pesticides or their separate needs were too small to justify the effort required to reach them. Moreover, the minimum purchasable amount was frequently larger than they could economically use.

Labor

The number of persons actively engaged in agriculture was virtually unchanged in the 1950's.[73] The reason is that employment opportunities on the farms did not keep pace with the natural increase in the rural population, so that a part of the farm labor force was forced to seek work in urban areas, or in nonfarm occupations in the countryside. Considerable evidence will be presented in Chapter 4 suggesting that Chilean farmers in the 1950's could have profitably used a substantially greater quantity of nonlabor inputs than were in fact employed and that had they done so the number of farm jobs available would have been much larger than it was. Had this occurred, the movement of workers from rural to urban areas doubtless would have been slowed and the supply of labor available to farmers would have been correspondingly greater. In other words, the failure of farm employment to grow in the 1950's did not reflect conditions on the supply side of the labor market.

The supply of labor is not just a function of the number of bodies available for work, however. It also depends upon the *quality* of the labor force. This is of particular importance in the process of agricultural development because the most efficient use of growing quantities of agricultural machinery, fertilizers, and pesticides requires a corresponding increase in the skills of persons working on farms. Hence it is important to consider the institutional mechanisms governing the rate of growth of skills relevant to the employment of these inputs.

While literacy is not essential to operation of a tractor or application of fertilizers and pesticides, it is surely very likely that knowledge of these skills can be more readily transmitted to a literate than to an illiterate population. Table 6 indicates that 36 per cent of Chile's rural population of working-force age was illiterate in 1952. In 1960 the percentage was 32 per cent. In contrast, the urban population of working-force age was only 10 per cent illit-

---

[72] Ibid., p. 60.
[73] The figures are given in the following chapter.

Table 6. Population of Chile According to Literacy, 1952 and 1960

	(Thousand persons over 5 years)			
	6 to 14 years		More than 14 years	
	1952	1960	1952	1960
Total urban and rural	1,252	1,601	3,682	4,442
Literate	773	1,145	2,954	3,724
Illiterate	479	456	728	718
Per cent illiterate	38.3%	28.5%	19.8%	16.2%
Total rural	566	585	1,369	1,393
Literate	265	341	876	947
Illiterate	301	244	493	446
Per cent illiterate	53.2%	41.7%	36.0%	32.0%

Sources:
1952–Servicio Nacional de Estadística y Censos, *Algúnos resultados del XII Censo de población y I de vivienda obtenidos por muestreo* (Santiago, no date).

1960–Dirección de Estadística y Censos, *Algúnos resultados provinciales del XIII Censo de población obtenidos por muestreo* (May 1963). The estimates are subject to a percentage error which diminishes with the size of the estimate. These errors range from 4 per cent for the estimate of rural illiterates ages 6 to 14 to 0.6 per cent for the estimate of total population over 14.

erate in 1952, and 9 per cent in 1960.[74] Evidently educational resources were considerably more abundant, or more efficiently used, in the urban areas than in the rural areas.

Since the 1920's Chile has had a compulsory education law for all children between the ages of 7 and 15. However, the law has not been observed in fact, as many children in this age group do not attend school. One indication of this is that between 1950 and 1959, of 1,966,300 children leaving primary school, only 28.6 per cent completed the full 6 years. Slightly more than 42 per cent completed only 1 or 2 years and were considered potential illiterates, and 28.9 per cent completed 3 to 5 years.[75]

There is strong evidence that these "desertion" rates are considerably higher in rural than in urban areas. This is indicated in Table 7, which is taken from a careful study of school attendance in Santiago Province. The data indicate that in rural areas and in towns of less than 5,000 population, 45 per cent of students enrolled two years previously in the first grade had left school by the beginning of the third year of study. In the city of Santiago and other towns larger than 5,000 population, about one-third of the students left by the beginning of the third year. In the rural areas, only 4.7 per cent of the

[74] These figures are implicit in Table 6.

[75] Universidad de Chile, Instituto de Organización y Administración, *Estudio de recursos humanos de nivel universitario en Chile* (Santiago, 1962), p. 21.

Table 7. Percentage of Students Remaining in School in Santiago Province

| School year | Rural areas | In towns with population of – | | City of Santiago |
		1,000 to 5,000	More than 5,000	
1	100.0	100.0	100.0	100.0
2	69.9	68.7	75.0	78.0
3	54.6	55.1	64.8	68.8
4	36.0	40.1	53.1	59.1
5	22.5	28.4	42.5	50.1
6	13.5	17.8	32.7	41.1
7	4.7	7.9	17.1	26.7

Source: Eduardo Hamuy, El problema educacional del pueblo de Chile (Santiago: Editorial del Pacífico, 1961), p. 86. The data refer to the 1950's.

original enrollment went beyond the full 6 years of primary study, and in towns of 1,000 to 5,000 population the figure was 8 per cent. By contrast, in Santiago 27 per cent of those who started primary school continued their education at the secondary level.

In this same study, data were examined for Curicó and Concepción Provinces, two of the principal agricultural areas south of Santiago, and the same pattern was revealed, except that the percentages of school leavers in rural areas were higher than in Santiago. In Curicó, 55 per cent left before the beginning of the third year and only 3.4 per cent went beyond the full 6 years of primary school. In Concepción, 52 per cent left before the beginning of the third year and only 3.1 per cent went on to secondary schools.[76]

The study of Santiago Province also considered the relationship between school leaving and income level in rural and urban areas. Three income levels were distinguished: high, medium, and low.[77] In the rural areas 97.6 per cent of the students were from low-income families, 2.4 per cent from medium-income families. There were no rural students classified in the high-income group.[78] Since it is quite obvious from common observation that there are high-income families living in rural areas, the implication of this finding is that the children from these families were sent to school in urban areas.

The findings showed that about 46 per cent of the low-income rural children did not get beyond the second year of school and only 5.6 per cent completed the full 6-year primary course of study. Among medium-income rural children 33.6 per cent dropped out by the end of the second year, and 14.4 per cent completed the full 6-year course.

[76]Eduardo Hamuy, El problema educacional del pueblo de Chile (Santiago, Editorial del Pacifico, 1961), pp. 87–88.

[77]The absolute income levels corresponding to each category were not given.

[78]Ibid., p. 112.

The following quotation, although from a study made in 1964, conveys an idea of the quality of primary education available in the 1950's to the relatively few rural children fortunate enough to finish the course of study:

> Of approximately 3,200 rural primary schools, 1,140 have only one teacher for all six grades and most of these teachers have only a primary education. Irregular student attendance (caused by economic factors, distance to schools, lack of motivation, parental apathy), lack of facilities, equipment and supplies, and lack of suitable financial rewards for teachers all combine to create a difficult situation.[79]

There is no doubt that the general educational level of the rural population of Chile in the 1950's was quite low. It is not surprising, therefore, that the percentage of farm people with specialized skills was correspondingly low. A CORFO report on manpower training in Chile asserts that plans for the mechanization of agriculture —

> encountered problems stemming from the lack of personnel trained in the operation and especially the maintenance of imported farm machinery. When acquiring machinery, farmers, industrialists, and construction firms had to budget substantial sums for maintenance, since rural and even urban workers were not trained to operate it properly. The consequent process of learning and experimentation, undertaken without basic mechanical knowledge, shortened the life of the machinery and entailed substantial expenditures on the purchase of spare parts.[80]

It was because of these deficiencies in mechanical skills that CORFO, in cooperation with the army, set up the already mentioned program for training men in the operation and maintenance of tractors and other farm machinery. Between 1942 and 1963, 6,551 persons graduated from this course, 1,943 of them being civilians who were allowed to attend the course at the request of the owners of farms where they were employed.[81] There were two kinds of instruction. One, lasting 5 to 7 months, was for conscripts and especially stressed machinery maintenance. The second was to train instructors for this (and other) courses, and it lasted 6 months. Between 1942 and 1963, 195 persons passed through this second type of course.

In 1954 the Institute for Rural Education (IER) was established with the objective of promoting improved levels of living for Chilean farm workers.[82] To achieve this objective the Institute put its major emphasis on education

---

[79]William J. Platt, Raymond A. San Giovanni, G. Allen Sherman, and Lloyd Dowler, *Training and Educational Needs in Chile's Agricultural Development*, Stanford Research Institute Project XI-9245 (Sacramento, Calif. June 1965), p. 9.

[80]CORFO, *La capacitación de mano de obra en Chile* (Santiago, 1965), p. 25. (My translation.)

[81]Ibid., pp. 26–27.

[82]Ibid., p. 52.

and training. It set up 21 Centers for Training of Farm Workers, each with a capacity of 50 students. Sixteen of these centers were for general orientation. They took farm workers between the ages of 16 and 30 and gave them, without charge, an intensive 4½ month course in general education, plus the rudiments of skills in fructiculture, horticulture, machinery operation and maintenance, blacksmithing, carpentry, beekeeping, dressmaking, weaving, canning, and home economics. They also attempted to encourage enthusiasm for rural cooperatives and community development in general.[83]

Two of the centers were for more advanced training, one in machinery operation and maintenance and the other in carpentry.[84] These courses also were for 4½ months. Students had to be graduates of one of the orientation centers, be at least 20 years old, and have completed the sixth year of primary school or its equivalent.[85]

One center was established in 1963 for the training of farm workers capable of managing and working properties taken over under the agrarian reform law of 1962. Students in this course had to be at least 20 years old and come from small-holder families or have experience in agriculture.

Finally, two of the centers were to train graduates of the orientation centers who had shown leadership qualities. The courses of these centers lasted one year, and their graduates worked with the Institute in a variety of programs devoted to rural community development: organization of centers for cultural and athletic activities, neighborhood and cooperative groups pursuing a variety of interests, and so on.[86]

Between 1955 and 1963, 8,212 students passed through the various courses offered by the Institute for Rural Education. Of these, 2,566 were graduated by 1960.[87]

In the period up to 1960 the only two programs of any importance for the training of agricultural technicians were those of CORFO-Army and the Institute for Rural Education. Through 1960, 5,919 persons had been through the CORFO-Army program[88] while, as just noted, the Institute had produced some 2,566 graduates. Thus the two programs had trained a total of 8,485 people in a variety of agricultural skills, the majority of them relating to the operation and maintenance of farm machinery. However, it was indicated in the earlier section dealing with machinery that a substantial number of the graduates of the CORFO-Army program entered nonagricultural occupations. Hence the number of specially trained persons actually working in agriculture in the 1950's must have been considerably smaller than the figure of 8,485 would suggest. Since there were some 650,000 persons occupied in the agri-

---

[83] Ibid., p. 57.

[84] These two centers were not established until 1962 and 1963 respectively.

[85] Ibid., p. 57.

[86] Ibid., p. 58.

[87] Ibid., p. 60.

[88] Ibid., p. 28.

cultural sector,[89] the conclusion is inescapable that in the 1950's only a very small proportion of the agricultural work force had special training and skills in the employment of such modern inputs as machinery, fertilizers, and pesticides.

The discussion so far indicates a very low level of general education in the rural education worked in such fashion as to impede the more rapid transformation of Chilean agriculture. This may appear paradoxical at first, since the to university-level training the situation is more difficult to evaluate. It is estimated that in 1961 there were 1,495 persons with university degrees in agriculture working professionally in that field.[90] Somewhat over half of these were trained at Catholic University and the remainder at the University of Chile. Complete information on their employment is not available, but data for 566 of them show that in May 1962, 385, or 68 per cent, were employed by the Ministry of Agriculture. Another 8 per cent were working for other government agencies.[91]

It is quite impossible to say whether the number of persons with university degrees in agriculture was "enough" in the 1950's, since there is no way of determining whether this factor of production was the element limiting the expansion of agricultural output. The number of such persons per head of population active in agriculture in 1960 was considerably less than in developed countries, but it was considerably more than in any other Latin American country except Costa Rica.[92] Mexico, for example, had only about one-third as many such persons per head of population economically active in agriculture. Yet in Mexico in the 1950's agricultural production grew much more rapidly than in Chile. Obviously there is no simple relationship between the rate of growth in agricultural output and number of university-trained people working in that field.

In general, it appears that the capacity of most Chilean farmers to make efficient use of modern inputs did not increase very much in the 1950's. Clearly the institutional mechanisms controlling the allocation of resources to rural education worked in such fashion as to impede the more rapid transformation of Chilean agriculture. This may appear paradoxical at first, since the allocation of resources to education occurs for the most part through political processes and the political clout of large landowners in Chile in the 1950's was considerable. The paradox is more apparent than real, however, since the large landowners typically educated their children in the cities and hence felt little personal incentive to improve the quality of rural schools. Moreover, one suspects that many, if not most, of the large landowners may have viewed better education for the mass of farm people as a potentially

[89] See Table 14.
[90] *Estudio de recursos humanos de nivel universitario en Chile*, p. 164.
[91] Ibid., p. 170.
[92] Ibid., p. 168.

disruptive influence, dangerous to the maintenance of the status quo in the countryside. Whatever the weight given to this explanation, the fact is undeniable that rural education in the 1950's was not conducive to the modernization of agriculture.

Water

About 50 per cent of total agricultural production in Chile in the 1950's was on irrigated land.[93] Hence, the conditions of supply of irrigation water obviously are of considerable importance in assessing the performance of agriculture. Farmers acquire water rights originally through grants from the state, while the actual day-to-day management of the water distribution system is in the hands of private farmer associations. Apparently there is no market in water rights apart from the land, i.e., the transfer of rights between farmers occurs when there is a transfer of land ownership.

It is widely held that the water distribution system in Chile is inefficient. Wollman quotes a joint IBRD-FAO report:

Most of the water comes directly from the rivers, whose flow is not regulated in any way. Thus coarse silt, gravel and even small boulders cannot be prevented from entering the canals. When floods occur, the head of the canal may be washed away and irrigation interrupted. Most canals are not protected against erosion and the water level decreases gradually; difficulties experienced in serving the fringes of the irrigated zone increase ... many small canals of great length were built by individuals or small groups with consequent unnecessary losses of water. Further lack of attention to irrigation during the night and weekends, except in the North where water is so precious that storage tanks are being built, causes additional losses. In the Central Valley fields with excessive applications of water (even to the point where crops suffer) alternate with others which get water only in a haphazard way.
... A conservative estimate puts the loss of water, i.e., the possibility of increasing the area under crops or artificial pasture, at 20 per cent.[94]

Apart from the efficiency of water distribution and use, and given the amount of precipitation, the supply of irrigation water depends upon the scale of investments in water storage facilities and in the network of canals for delivering water to the farm. While many of the early irrigation works were privately built, all those constructed in the last several decades have been government projects. A number of these were undertaken in the 1950's — although, as indicated in the quotation above, the rivers in Chile

[93] Information provided by the Agricultural Planning Office (ODEPA).

[94] Joint Mission of the International Bank for Reconstruction and Development and the Food and Agriculture Organization of the United Nations, *The Agricultural Economy of Chile* (Washington, 1952), pp. 176 ff., quoted in Nathaniel Wollman, *The Water Resources of Chile* (Baltimore: The Johns Hopkins Press for Resources for the Future, 1968), pp. 43–44.

remained essentially unregulated — but it appears that actual additions to the irrigated area in the decade were quite small.[95]

Hence the supply of irrigation water appears not to have been significantly increased in the 1950's, and the system for distributing it suffered from serious inefficiencies. Yet it does not appear that water supply was an important factor in limiting the expansion of output over the decade. The argument in support of this conclusion is presented in Chapter 4. Suffice it to say here that there is considerable evidence showing that given the supply of fertilizers, pesticides, machinery, and other nonland inputs, it would have been uneconomical to bring more land under irrigation. In this important economic sense, irrigated land was not in short supply and hence did not limit the expansion of output.

Land

The kinds and amounts of land available to any country are fixed over any period of time useful for economic analysis. The uses made of the land, and the kinds and amounts of other inputs combined with it, are not fixed, however, and variations in these factors will explain variations in the yield of the land. Hence, in the consideration of the institutional conditions which control the "supply" of land, attention should be focused on the way in which these conditions affect land management and the incentives and abilities of landowners to adopt improved inputs.

It is widely held that the land tenure system in Chile has resulted in a generally inefficient use of land, water, labor, and capital and has retarded the rate of adoption of new, more productive inputs. This is a theme that runs through most of the discussions of Chilean agriculture that have appeared over the last few decades, and it has provided the motivation and the argument for the various land reform measures adopted in this period. One of the most complete and clear presentations of this point of view appears in a report which deals specifically with the relationship between Chile's land tenure system and the rate of socio-economic development of the agricultural sector.[96]   There it is asserted that —

> Almost nowhere is full advantage being taken of land and water resources, but this is especially notable on large farms. In addition there is considerable underemployment of labor, especially among those families which depend on the operation of sub-family-size holdings. However, underemployment is evident also on the large farms. The amount of capital invested is relatively low and much of it is inefficiently used. Both case

---

[95] The evidence on this is considered in the next chapter.

[96] CIDA Report. The Chilean agricultural plan, published after this was written, presents a diagnosis of the country's agriculture which on essential points is the same as that of the CIDA Report. See República de Chile, Ministerio de Agricultura, *Plan de desarrollo agropecuario 1965–1980*, 4 vols. (Santiago, 1968).

studies and data for the country as a whole indicate that this failure to take full advantage of available economic resources is correlated with the traditional system of *latifundio-minifundio.*[97]

It is noted that the *minifundistas* constitute about 70 per cent of all farm families, and that they have no chance to get land and capital on the scale necessary to produce more efficiently. In addition, they have inadequate access to credit, irrigation water, technical assistance, and product markets. Only half of their children of school age actually attend school, and it is rare for one of them to finish high school.[98]

The argument asserts further that the *latifundio-minifundio* system discourages adoption of improved inputs and farm practices. The small holders lack access to these inputs and the technical knowledge needed to use them efficiently, while sharecroppers and farm workers resident on the *latifundio* lack the incentives. Sharecropper and employment contracts are generally verbal and can be ended at the will of the *latifundista.*[99] He must give his approval for any improvements the sharecropper may wish to make, and the latter has no assurance that he will receive the benefits of such improvements.[100]

The system also provides no incentives to the *latifundista* to employ his resources efficiently or to readily adopt new inputs and practices. Even though he has much the greater part of agricultural credit, ready access to the best land, a relatively abundant supply of water and labor, and easy access to urban markets and technical knowledge, he is under neither economic nor social compulsion to make full use of these advantages. By virtue of his position as a large landowner he enjoys both social status and political power. He receives an income far above that of the *minifundistas*, even though he does not work his land intensively. Besides, he frequently has business, professional, or political interests in the cities so that he is not dependent upon agriculture as his only source of income. For all these reasons he is under little if any pressure to produce efficiently or innovate.[101]

---

[97]Ibid., p. 205. This and all subsequent quotations from the CIDA Report were translated by the present author. The principal characteristic of the *latifundio-minifundio* system is a highly unequal distribution of land, which produces correspondingly unequal distribution of economic, political, and social power between the *latifundistas* and *minifundistas*. The latter constitute the bulk of the labor force employed by the former. They are small farmers, or members of their families, whose holdings are not large enough to provide them full-time employment, sharecroppers, and workers living on and employed by the *latifundio*, who receive the right to work a small piece of land as part of their compensation.

[98]Ibid.

[99]This was true in the 1950's. Legislation passed after 1964 strengthened considerably the job tenure situation of farm workers.

[100]CIDA Report, pp. 205–06.

[101]Ibid., p. 206.

Thus runs the argument. There cannot be the slightest doubt concerning the validity of much of it. The evidence is overwhelming that the distribution of resources in Chilean agriculture is extraordinarily unequal. Table 8 demonstrates this with respect to land. It shows that 78.6 per cent of all agricultural land, 65 per cent of arable land, and 78 per cent of irrigated land was found on but 7 per cent of the farms.

Table 8. Number and Area of Farms Classified by Size, 1955

| | Subfamily | | Family | | Multifamily | | | |
| | | | | | Medium | | Large | |
	Number or area	Per cent	Number or area	Per cent	Number or area	Per cent	Number or area	Per cent
Number of farms	55,761	37.0	60,388	40.0	24,427	16.0	10,383	7.0
Area (thous. hectares):								
Agricultural	67.4	0.3	1,767.8	8.1	2,823.0	13.0	16,983.9	78.6
Arable	57.5	1.0	642.6	12.0	1,220.4	22.0	3,623.0	65.0
Irrigated	23.4	2.1	80.0	7.3	138.3	12.6	856.2	78.0

Source: CIDA Report, p. 168, based on the agricultural census of 1955.

Note: Farms are classified in one of the four categories depending upon size and the number of man-years needed to work them. Any farm which requires less than 2 man-years of labor is classified as subfamily, and all farms of less than 5 hectares are in this category. Family farms are those which require 2 to 4 man-years. Their size varies from 5 to 19.9 hectares in the northernmost areas and from 5 to 1,999.9 hectares in the extreme south. Medium-sized multifamily farms require 4 to 12 man-years and vary from 20 to 199.9 hectares in the north to 2,000 to 4,999.9 hectares in the south. Large multifamily farms require more than 12 man-years and vary in size from 200 or more hectares from the northernmost province down through Cautín to 5,000 or more hectares in the far south. CIDA Report, p. 283.

The concentration of credit in the hands of the large farmers has already been noted in an earlier section, and the strong relation between credit availability and the purchase of farm machinery leaves little doubt that small farmers have had only very limited opportunities to mechanize.[102] The crucial role of credit also extends to fertilizers and pesticides, with the same unfavorable results for small farmers. Finally, it is only the sons of larger farmers who take university degrees in agricultural subjects, or otherwise acquire technical training. As noted earlier, only a very small proportion of the children of small farmers even finish primary school. Thus knowledge, perhaps the most important resource of all, is also very unequally distributed.

[102] Evidence is presented in Chapter 5 indicating that about 100 to 200 hectares was the minimum size of farm consistent with efficient use of a tractor and attendant equipment. This size limitation would not apply to other simpler kinds of machinery, however, the use of which was denied to small farmers, in part at least because of inadequate credit.

The unequal distribution of resources has resulted in a similarly unequal distribution of income and of social and political power. Consequently the vast majority of farm people have traditionally lived poorly and have been unable to exert enough influence to protect and advance their interests. Thus the generally low level of public services, and above all of educational services, in the countryside is linked to the unequal distribution of private resources, especially land.

It seems fair to say that the land tenure system in Chile, coupled with the operation of credit institutions, has made it virtually impossible for the great bulk of the farm population to acquire the knowledge and other inputs essential to a modern, technically evolving agriculture. This very important part of the argument typified by the CIDA Report seems irrefutable.

Other aspects of the argument are less convincing, however. It is asserted that resources generally were inefficiently used in Chilean agriculture, and that the spirit of technical innovation was weak among large landowners. Both of these faults are attributed directly to the system of land tenure. Much is made of the fact that a seemingly large proportion of both arable and irrigated land was in natural pasture. In 1955 almost 53 per cent of total arable land was devoted to this use; for irrigated land the proportion was 29 per cent.[103] It is asserted that these percentages reflect an "enormous waste" of land resources; however, the only basis for this judgment is that this extensive use of land persisted in a period when agricultural imports were increasing. The logic seems to be that rising agricultural imports indicated increasing domestic demand that could have been economically satisfied by cultivating the land more intensively. However, the mere coexistence of increasing farm imports and extensively used land does not constitute proof that the land was being used inefficiently. The question is whether, given the costs and returns of all resources employed, or employable, in agriculture, it would have been economical for farmers to utilize the land more intensively. To do so would have required substantially more nonland inputs. The CIDA Report argument implicitly assumes that these greater quantities of inputs were in fact available at economical prices but that farmers simply were not interested in using them. This underlying assumption is never brought to the fore, however, in the discussion of the efficiency of land use. There is no systematic analysis of the terms on which these inputs were available and hence no basis for judging whether farmers in fact had both opportunities and incentives to use more of them.[104]

---

[103]CIDA Report, pp. 146 and 148.

[104]Input-output price relations are treated quite superficially in another context on pp. 245 and 246 of the CIDA Report. Most of the comparisons are between agricultural commodity prices and those of all nonagricultural commodities, industrial products, and imported products. The point of these comparisons is not obvious. The relation of agricultural input prices to output prices is dealt with in one paragraph. The different movements in prices of agricultural labor and other inputs, noted on pp. 9–11 above, are not considered.

The CIDA Report considers the pattern of land use in each of four size classifications and concludes that in each group there was considerable under-utilization of both arable and irrigated land.[105] The data presented, however, show that very large multifamily farms used land slightly more intensively and medium-sized multifamily farms slightly less intensively than family-sized farms. The only clear difference between size groups with respect to intensity of land use was between these three groups on the one hand and the sub-family-sized farms on the other.[106] As far as it goes, this scarcely suggests a continuous relationship between efficiency of land use and size. But for the reasons already noted these comparisons of percentages of land in various uses tell very little about efficiency.

The CIDA Report asserts that there is very substantial unemployment of agricultural labor, and that this is due in large part to the system of land tenure and especially to size of farm.[107] The evidence presented on the amount of unemployment indicates that about one-third of the farm labor force was surplus in 1955; that is, the output achieved that year could have been produced with only about two-thirds of the actual agricultural labor force. This estimate is based on the number of man-days per year needed to cultivate one hectare of each different crop, multiplied by the number of hectares devoted to each crop. The estimate of total man-days needed is then divided by 300, the number of days assumed to represent full employment.

The estimates of man-days needed per hectare of each crop assume "simple mechanization," although this phrase is never defined. No account is taken of seasonal demands for labor, undoubtedly an important factor because the size of the permanent work force employed in agriculture depends on them as well as on the area and crop pattern of cultivation. Suppose, for example, that there is only one crop, that each hectare requires 60 man-days per year of labor, and that there are 10,000 hectares planted to this crop. Thus the total demand for labor is 600,000 man-days annually. If there is no seasonal pattern in the demand for this labor and if 300 working days per year per man corresponds to full employment, then the full employment size of the labor force is 2,000 men. But suppose that the demand for labor is seasonal, reaching a peak in the month of harvest when 90,000 man-days are needed. The 2,000 men working all day every day in the month could provide only 60,000 man-days. Unless labor can be attracted from nonagricultural sectors to cover seasonal demands, the 2,000-man permanent work force will not be big enough. Under this assumption a work force of 3,000 men would be necessary to satisfy the total demand for labor, given its seasonal pattern. But since the annual demand for labor is only 600,000 man-days, the average work year per man will be only 200 days. By the criteria of the CIDA Report, there would be severe underemployment. This is strictly seasonal under-

[105] Ibid., p. 149.
[106] Ibid., pp. 147–48.
[107] Ibid., p. 151.

employment, however, reflecting particular technical and labor market conditions. It is not general underemployment as that term is typically used, nor does it reflect inefficiency in the employment of labor from the standpoint of individual farm employers. To be sure, if the difference between seasonal peak and trough demands for labor is very large, then from the standpoint of society as a whole it may pay to seek means of reducing it by improving storage and transport facilities and promoting greater regional and sectoral mobility of labor. But these are not remedies that can be effected through the initiative of individual farmers or of the agricultural sector alone. Hence, in judging the efficiency of resource use within that sector, the technical and market conditions which determine the extent of seasonality in the demand for labor must be taken as given. In this case, the greater the difference between seasonal peak and trough demands, the smaller the number of annual man-days per man consistent with full employment.

There is considerable seasonality in the demand for labor in Chilean agriculture. Hence the CIDA Report's failure to allow for seasonal fluctuations means that its estimate of the number of workers needed probably is too low and its estimate of unemployment, therefore, too high.[108]

There are additional grounds for thinking that the CIDA Report's estimate of underemployment of labor is too high. Apparently no allowance is made for labor time devoted to work other than cultivation: construction and maintenance of buildings, fences, and irrigation canals, for example.[109] Nor does it appear that account is taken of days lost to vacations, holidays, and bad weather.

Thus the assumption that 300 days per year is consistent with full employment is almost surely exaggerated. Consequently the estimate of labor surplus in the 1950's is probably too high.[110] Even allowing for a substantial error, however, it probably is true that there was some underemployment of the agricultural labor force in that period in the sense that some wage earners would have worked more at the prevailing wage rate had they had the opportunity. Moreover, there probably were small self-employed farmers who, while not willing to work more for the increment of income obtainable for additional effort, would have been willing to work more if more land and other inputs had been available to them, thus assuring a greater pay-off to additional work. While these people were not technically unemployed—they voluntarily decided not to work more—they could be so regarded from a broader social standpoint. It may be granted, therefore, that there was under-

[108] The report says that the failure to allow for seasonality is compensated for in part by the assumption of "simple mechanization" in techniques used. However, no explanation is given of how this compensation occurs, and it is not obvious on its face.

[109] See p. 27, n. 15, of the report, where estimates of required man-days are given only for various crops and animals.

[110] In the agricultural plan it is estimated that farm unemployment was 19 per cent in 1965 (*Plan de desarrollo agropecuario*, Vol. II, p. V-116).

utilization of labor in the 1950's. Whether this was attributable primarily to the land tenure system, however, is another question, treated below.

The Report states that both physical capital and technical knowledge were wastefully used in agriculture.[111] Some of the material presented in support of this argument is irrelevant; for example, a comparison between Chile and certain other Latin American countries as to the ratio of gross fixed investment to gross national product. Other evidence is more persuasive, however, such as the generally poor maintenance of irrigation canals, high losses of animals to preventable diseases, and generally poor livestock management practices. All of this is quite impressionistic, however, since the standards of efficiency applied are vague or not evident at all. No analysis is made of the costs involved in adopting what appear to be more efficient practices.

Some attention is given to the utilization of capital on farms of different size. The available data are very meager, however, and it is concluded that there is no discernible relation between size and efficiency of capital utilization.[112]

It is almost certainly true that the land tenure system slowed the rate of adoption of new production techniques, since it consistently denied the great bulk of the agricultural labor force access to education and other modern inputs. It is not so clear, however, that the system retarded the rate of innovation to the extent suggested by the CIDA Report. The difference in emphasis lies in the assessment and explanation of the performance of larger farmers. As noted above, the CIDA Report argues that the spirit of innovation was very weak among large farmers because they were not much interested in additional income. Several sorts of evidence are presented in support of this argument. One is an article by Kaldor demonstrating that the share of profits in Chilean national income is very high compared with that in developed countries but the saving rate is very low. The implication is that the capitalist class saves a relatively small proportion of its income.[113] Although Kaldor's article is based on analysis of the Chilean national accounts, the CIDA Report concludes that the results are *especially* applicable to agriculture because that has been the least dynamic sector of the economy.[114]

Additional evidence is drawn from a study by Sternberg of the saving and consumption patterns of 20 large landowners in the Central Valley.[115] These 20 farmers had an average disposable personal income of E°42,535 (about US$40,000) in 1960. Of this they consumed 83.7 per cent. This figure seems high and it is consistent with Kaldor's findings for the economy as a whole.

---

[111] CIDA Report, p. 157.

[112] Ibid., pp. 157–58.

[113] Nicholas Kaldor, "Problemas económicos de Chile," *El trimestre económico*, 26 (1959), 170–221.

[114] CIDA Report, pp. 184–85.

[115] Marvin J. Sternberg, "Chilean Land Tenure and Land Reform," Ph.D. thesis, University of California, 1962.

Whether it is typical of large landowners generally is uncertain, however. A sample of only 20 farms suggests that the results may not be representative of the larger universe from which the sample was taken. The CIDA Report, for example, studied 10 large landowners in Valdivia Province and discovered that 35 per cent of their income was reinvested in their farms.[116] This is clearly inconsistent with the Sternberg result. Thus while there is evidence to support the argument that large landowners save relatively little, and therefore may lack the spirit of innovation, the evidence is not conclusive.

Even if it were, the conclusion would not follow that the land tenure system was responsible for the failure of large landowners to invest more. Kaldor's study indicates that the saving rate for the whole economy was low. The share of agricultural income in national income was not large enough to explain this result. Conceivably the agricultural saving rate was lower than that of other sectors—although this is not proved—but the rate in these other sectors must also have been relatively low. This suggests some factor which acted generally to discourage saving. The most obvious candidate for this role is the rampant inflation which has existed in Chile almost without interruption since before the end of the last century. Such inflation would surely discourage liquid saving and thus make more difficult the development of those financial intermediaries which play such an important role in mobilizing liquid savings for investment. Moreover, rapid inflation would probably also discourage direct investment in assets which pay off over a long period of time and for which there is not an active market. Land improvements are an example. Hence the existence of a low savings rate among large landowners, assuming that it was in fact low, was not necessarily a result of the land tenure system. The inflation explanation is at least as plausible, if not more so.

In further support of its argument that large landowners lack the spirit of innovation, the CIDA Report alleges that they were not responsive to price changes. Several studies of the response of cultivated area to relative price changes are cited in support. One such study showed that for the country as a whole a 10 per cent increase in the relative price of wheat produced a 2.5 per cent increase in the area sown to wheat.[117] Another study of the southern area of the country showed that a 10 per cent increase in the relative price of wheat produced a 4 per cent increase in the area sown. This study showed similar results for onions, garlic, and rice, whereas for sugar beets and sunflower seed, the percentage responses in area sown exceeded the percentage increase in relative prices.[118]

---

[116]CIDA Report, p. 188.

[117]César Carmona and J. Luis Marambio, "Comercialización del trigo en Chile," thesis presented as part of the requirements for the degree of agricultural economist, University of Chile, 1964.

[118]Roberto Echeverría, *Respuesta de los productores agrícolas ante cambios en los precios, Informe de Avance de la Investigación.* (Instituto de Capacitación e Investigación en Reforma Agraria en colaboración con el Land Tenure Center de la Universidad de Wisconsin, [1965]).

The CIDA Report asserts that the elasticities found for wheat, onions, garlic, and rice are low, i.e., that they indicate a lack of responsiveness by farmers to relative price changes. In fact these elasticities are comparable to those found in other countries, including the United States. One study, for example, showed the following elasticities in the United States:[119]

	Percentage response of area sown to 10
Crop	*per cent change in relative price*
Cotton	3.4
Wheat	4.8
Corn	1.0

A study of supply elasticities for various crops in India found that they ranged from 0.8 per cent for irrigated wheat (i.e., a 10 per cent change in price yielded a 0.8 per cent increase in wheat acreage) to 7.2 per cent for cotton.[120] The responsiveness of Chilean farmers to price changes, therefore, appears to be in line with that of farmers in other countries, including the most developed.

This conclusion is reinforced, in fact, by material presented in the CIDA Report. In discussing the high response to changes in relative prices of sugar beets and oil seed crops, the Report says:

> The elastic response of producers of sugar beets and sunflower seed was associated not only with an increase in prices but also with the circumstance that they were new crops for Chile and that they were favorably priced in relation to traditional crops, which produced a reduction in the area sown to the latter.
>
> For example, the rapid expansion of the area sown with oil seeds has been in large part at the expense of the area previously sown with wheat.[121]

This statement indicates that because oil seed prices were favorable in relation to wheat prices, Chilean farmers switched from wheat to oil seeds, i.e., they were responsive to relative prices. The statement also reveals a weakness in the previously cited studies of wheat area responses. Those studies measured the response of wheat area to changes in the price of wheat in relation to the wholesale price index of agricultural commodities. Yet the statement just quoted indicates that the price of oil seeds was the more relevant variable in determining changes in wheat area. If this were true, then

[119] Marc Nerlove, "Estimates of the Elasticities of Supply of Selected Agricultural Commodities," *Journal of Farm Economics*, 38 (1956), pp. 496–509.

[120] R. Krishna, "Farm Supply Response in India-Pakistan: A Case Study of the Punjab Region," *Economic Journal*, Sept. 1963, pp. 477–87.

[121] CIDA Report, p. 209.

a study relating wheat acreage to the price of wheat deflated by the price of oil seeds would show greater responsiveness than is found in the studies cited.

In short, the evidence presented by the CIDA Report concerning price responsiveness of Chilean farmers does not support the view that they respond sluggishly. Quite the contrary. It indicates that Chilean farmers behave much like farmers in other countries of the world, including those with a highly developed agriculture.

Finally, it is hard to square the CIDA Report argument concerning large farmers with the fact that in the 1950's there was a considerable increase in the use of machinery, fertilizers, and pesticides, the inputs required to shift from traditional to modern agriculture.[122] Since the increased use of these inputs was financed to a large extent by credit, and since credit was heavily concentrated in the hands of larger farmers, the conclusion seems inescapable that they were responsible for the rather considerable rate of innovation observed in the 1950's.

Data from the 1955 and 1965 agricultural censuses strengthen the impression that the larger farmers in Chile were technologically more advanced than the smaller farmers. In both years yields per hectare for most crops were higher on large than on small farms, yields generally increased more in percentage terms on large than on small farms from 1955 to 1965, and larger farmers fertilized a greater proportion of their land in crops and cultivated pasture than smaller farmers. (See Tables 9 and 10.) None of this is particularly surprising, given the concentration of resources in the hands of the larger farmers. It is worth noting, however, because of its bearing on the CIDA Report's argument that the larger farmers were generally uninterested in technological advance.[123]

The weakness of the CIDA Report's argument regarding larger farmers has a bearing on that part of the analysis which attributes the underemployment of labor to the land tenure system. It is asserted that under the system a very large percentage of farmers were denied sufficient land and other resources to provide them with full-time employment. That the distribution of land and other resources was unequal is indisputable. Whether the associated underemployment of labor was attributable to the tenure system is not so clear, however. The question naturally occurs, why did the underemployed farmers and farm workers not find employment on the larger farms where availability of land and other inputs apparently was no problem?

This question is not posed in the CIDA Report. However, the treatment of the efficiency of resource use on large farms indicates clearly what the answer would be: large farmers were content with what they had and simply were

---

[122]The material dealing with the increased employment of these inputs is presented in the next chapter.

[123]This argument is made in even stronger form in the agricultural plan for 1965–80. There it is asserted that the larger farmers are not only disinterested in technological improvements but actually less efficient than the smaller farmers.

Table 9. Output per Hectare of Selected Crops, 1955 and 1965

(Quintals)[a]

Farm size (hectares)	Wheat			Corn			Beans			Potatoes			Sugar beets		
	1955	1965	Per cent increase	1955	1965	Per cent increase	1955	1965	Per cent increase	1955	1965	Per cent increase	1955	1965	Per cent increase
0.1–5	10.6	10.1	−4.7	13.2	23.2	75.8	9.0	11.6	28.8	64.7	68.1	5.3	—	336.2	—
5.1–10	10.2	11.3	10.8	12.5	16.2	29.6	8.5	8.3	−2.4	67.4	56.5	−16.2	30.9	334.0	980.9
10.1–20	10.1	9.8	−3.0	11.9	16.8	41.2	7.7	7.7	0.0	61.2	60.0	−2.0	156.2	325.8	108.6
20.1–50	10.7	10.4	−2.8	11.0	20.7	88.2	7.6	7.2	−5.3	74.4	59.6	−19.9	176.1	359.8	104.3
50.1–100	11.7	12.7	8.5	12.1	29.0	139.7	8.2	8.1	−1.2	79.3	72.1	−9.1	164.2	356.5	117.1
100.1–200	13.8	15.6	13.0	15.0	35.4	136.0	9.8	9.7	−1.0	95.6	94.6	−1.0	142.6	361.7	153.6
200.1–500	14.5	18.5	27.6	17.6	37.0	110.2	9.9	11.0	11.1	105.3	106.5	1.1	212.4	401.3	88.9
500.1–1000	16.4	19.5	18.9	15.2	34.7	128.3	11.4	9.7	−14.9	103.4	108.0	4.4	264.3	410.2	55.2
1000.1–2000	15.2	20.1	32.2	16.7	36.3	117.4	10.4	11.0	5.8	98.2	102.4	4.2	237.8	358.5	50.8
2000.1–5000	15.3	18.6	21.6	19.9	34.2	72.9	10.4	11.6	11.5	108.7	111.2	2.3	117.3	358.4	205.5
+5000	12.6	18.7	48.4	18.9	35.7	88.9	11.9	14.5	21.8	110.8	107.3	−3.2	239.9	378.0	57.6
Country	13.9	15.5	11.5	15.4	29.6	92.2	9.8	10.1	3.0	87.6	77.2	−11.9	216.0	379.0	75.5

Source: Information provided by ODEPA based on 1955 and 1965 agricultural censuses. In both 1955 and 1965 the five commodities in the table accounted for about 80 per cent of total crop production, excluding fruits, vegetables, and wine grapes. For these three classes of commodities, which were about 40 per cent of total crop production in 1955 and 42 per cent in 1965, yields did not vary significantly with size of farm.

Excluding the two southernmost zones, which accounted for only about 18 per cent of total crop production in 1955, the size classes of farms were as follows: subfamily: 0.1–4.9 hectares; family: 5–19.9 hectares in the two northernmost zones and 5–49.9 hectares in the central zone; multifamily medium: 20–49.9 hectares in the two northernmost zones and 50–199.9 hectares in the central zone; multifamily large: over 200 hectares.

[a]1 quintal = 220.5 lbs.

Table 10. Per Cent of Land in Crops or Cultivated Pasture to Which Fertilizer Was Applied, by Farm Size

Farm size (hectares)	Per cent fertilized	
	1955	1965
0.1–5	13.8	12.7
5.1–10	22.8	38.5
10.1–20	24.5	39.6
20.1–50	24.8	43.2
50.1–100	29.7	49.8
100.1–200	30.7	54.6
200.1–500	33.0	52.5
500.1–1,000	36.6	49.5
1,000.1–2,000	36.0	47.4
2,000.1–5,000	34.8	41.7
+5,000	19.4	20.2

Source: ODEPA, based on agricultural censuses for 1955 and 1965. See Source, Table 9, for number of hectares in the four size classes of farms: subfamily, family, multifamily medium, and multifamily large.

not interested in producing more even though large supplies of cheap labor were at hand. Thus the argument that the land tenure system was responsible for the underemployment of labor rests ultimately on the presumed effects of the system in sapping the initiative of large landowners, making them insensitive to economic stimuli. The weakness of the evidence supporting this presumption and the abundance of evidence controverting it have already been considered. Hence the argument that underemployment of labor was attributable primarily to the land tenure system is on shaky ground.[124]

To recapitulate, there is much truth in the argument that the Chilean land tenure system retarded the rate of adoption of new agricultural inputs in the 1950's. The principal impact, however, seems to have been on the small farmers and farm workers—the great mass of farm people—who were deprived of adequate access to education and other modern inputs. The available evidence does not demonstrate that the land tenure system induced larger farmers to manage their resources inefficiently, or that it was responsible for underinvestment by them, or that these farmers were unresponsive to price and other economic stimuli.

[124]The question remains: why did the large landowners not hire greater numbers of the unemployed and underemployed farm workers? Considerable evidence is presented in Chapter 4 showing that Chilean farmers generally, including larger farmers, could have profitably employed substantially more labor and other nonland inputs, except livestock, in the 1950's and that, at least among the larger farmers, their failure to do so was attributable to government credit and import policies which held the supply of imported inputs below the demand for them. Had greater quantities of these inputs been available, the larger farmers would probably have purchased more of them and also have employed more labor.

Summary

This discussion of the supply conditions of agricultural inputs in the 1950's points up some potentially serious obstacles to the rapid modernization of farming. The level of general education and technical training of the great mass of the farm labor force was low and did not improve significantly over the decade. This must inevitably have slowed the rate of adoption of new inputs. Moreover, the sheer physical availability of these inputs was subject to a variety of import controls, the restrictive effects of which were reinforced, particularly in the second half of the decade, by a policy of credit stringency. Finally, the land tenure system must have seriously weakened the ability and willingness of most farmers to innovate.

With this sketch of the conditions of input supply in mind, let us turn now to consider the actual performance of agriculture in the 1950's.

CHAPTER

3

Growth of Production and Inputs in the 1950's

GROWTH OF AGRICULTURAL PRODUCTION

Agricultural production – crops and animal products – accounted for almost 15 per cent of the gross geographical product of Chile in the 1950's.[1] The sector declined somewhat in relative importance over the course of the decade, as production grew by slightly less than 2.5 per cent annually (see Table 11) whereas the country's gross geographical product grew at an average annual rate of 3.6 per cent.[2]

Agricultural production in Chile provides an excellent example of the dangers involved in aggregating diverse economic quantities as though they were homogeneous. Although, as indicated, total production in the 1950's grew by somewhat less than 2.5 per cent annually, this average conceals quite distinctive patterns in the growth rates of the principal commodity groups. Total crop production grew by 39 per cent, an average annual rate of 3.3 per cent; livestock production grew by 17 per cent, or 1.6 per cent annually; but

[1] Universidad de Chile, Instituto de Economía, *La economía de Chile en el período 1950-1963*, II: *Cuadros Estadísticos* (Santiago, 1963), Table 7.

[2] To avoid distortions arising from the use of single beginning and terminal years, these changes are calculated over the period 1949/1951 to 1959/1961. The figures for gross geographical product are based on *La economía de Chile,* II, Table 7. Those for agricultural production are based on indexes of the physical volume of production from Presidencia de la República, Oficina de Planificación Nacional, *Indice de producción agropecuaria-silvícola, 1939-1964* (mimeo., not paginated; Santiago, 1966).

In the national accounts of Chile there are estimates of "real" agricultural production, i.e., value of agricultural output expressed in constant prices. In principle these ought to move proportionally to the index of the volume of production, but in fact they do not. The reason is that the constant price estimates are derived by deflating the current value of output by the index of the general price level whereas the indexes in *Indice de producción* are based on the physical volume of output of each product weighted by the total value of its output in the base period. The *Indice de producción* figures are undoubtedly better measures of movements in the volume of production than those in the national accounts.

Table 11. Growth of Agricultural Production in Chile in the 1950's

	1949/51–1954/56		1954/56–1959/61		1949/51–1959/61	
	Per cent increase	Average annual Per cent increase	Per cent increase	Average annual Per cent increase	Per cent increase	Average annual Per cent increase
Total	13.6	2.60	11.8	2.25	27.0	2.45
Crop production	20.0	3.70	15.5	3.00	39.0	3.33
Animal products	6.7	1.30	9.6	1.85	17.0	1.60

Source: Indice de producción. Revised but still unpublished figures prepared by ODEPA indicate somewhat lower growth rates than those given in the table. The differences do not significantly affect any of the conclusions stated in the text, however.

production of fish and fish products declined by 35 per cent, an average annual rate of somewhat more than –4 per cent. Fish production is of slight importance compared to crops and livestock and can be ignored in this analysis.

It is commonly asserted that the expansion of Chilean agriculture has been "inadequate." The growth of agricultural production for all Latin American countries as a whole, for example, was 50 per cent in the 1950's compared with Chile's 27 per cent.[3] A more relevant point is that since agricultural production accounted for almost 15 per cent of gross geographical product in the 1950's, its growth rate necessarily had a significant effect on the rate of growth in total production. Perhaps of greater importance, the slow expansion of agricultural production and productivity was directly related to a sizable increase in food imports, since many of these imports were of commodities produced in Chile. Had productivity increased fast enough, domestic production would have been sufficient to satisfy the rising demand for food, thus reducing the need for imports without diverting resources from export industries. The amount of foreign exchange used to finance higher food imports thus would have been available for the purchase of capital goods. The slow growth of agricultural productivity and production, therefore, may have significantly reduced the rate of capital formation in the country.

There can be little doubt that the overall growth of the Chilean economy in the 1950's was retarded by the relatively poor performance of the agricultural sector. Yet the figures cited above indicate that the major culprit was livestock production. Had this kept pace with crop production the growth rate for the sector as a whole would have been 3.3 per cent annually rather than 2.5 per cent. In comparison with the rest of Latin America, Chile would still have been a laggard, but the discrepancy would have been much smaller.[4]

[3] Agricultural production in Latin America from United Nations, Economic Commission for Latin America, *Statistical Bulletin for Latin America*, LV, 1 (Santiago, Feb. 1967), p. 12. The comparison is between the years 1949/51 and 1959/61.

[4] Annual average growth in total agricultural production for all of Latin America was 4.1 per cent between 1949/51 and 1959/61. Ibid.

The improvement in the balance of payments would have been significant. If livestock production had grown by 3.3 per cent annually in the 1950's, it would have reached E°291 million in 1960. Actual production that year was E°245 million.[5] Had the higher growth rate been achieved, therefore, production in 1960 would have been about E°45 million more than it was in fact. At the then prevailing rate of exchange (E° 1 = $.95) this unrealized production would have amounted to some $42 million. Imports of animal products in 1960 were $37.8 million.[6] Of course it cannot be assumed that the increased production would have exactly offset these imports, since the product composition of the two totals would not have been precisely the same. However, given total domestic consumption of animal products in 1960, the higher production would either have displaced imports or have added to exports. In either case, the amount of foreign exchange attributable to animal production would have been greater by some $42 million. This is equal to 26 per cent of total imports of capital goods in 1960.[7]

There were significant differences in the growth rates of the several components of the animal products group. Output of poultry and products increased by 48 per cent in the 1950's, while meat production, aside from chickens, grew by only 9 per cent. Dairy products were higher by a little over 16 per cent, but production of wool was unchanged.[8]

There were also significant variations in the growth rates of the various components of crop production. This is clearly seen in Table 12.

GROWTH OF AGRICULTURAL INPUTS

The available data for agricultural inputs in Chile are such that a precise account of their growth over the decade of the 1950's is impossible. They are adequate, however, to support some reasonably well-grounded judgments.

Land

The relevant material is presented in Table 13. Curiously enough, there are major uncertainties concerning the total amount of land fit for agriculture in Chile, and the amount that was under irrigation in the 1950's. This is indi-

---

[5]The E°245.2 million figure, as noted earlier, was 17 per cent above production in 1949/51. Actual production in those years, therefore, averaged E°209.4 expressed in 1960 prices (E°245.2 ÷ 1.17). E°209.4 at 3.3 per cent annually for 10 years is E°291 million. Actual production in 1960 is from Ministerio de Agricultura, Departamento de Economía Agrícola, *La agricultura chilena en el quinquenio 1956-1960* (Santiago, 1963), p. 88.

[6]Ibid., p. 212.

[7]Capital goods imports in 1960 were $173 million. See *La economía de Chile,* II, p. 62.

[8]All figures from *Indice de producción.*

Table 12.   Growth Rates in Volume of Production of Various Crops and Their Relative Importance in Total Production

Kind of crop	Annual growth rate (%) 1949/51-1959/61	Percentage share of value of total crop production 1955/57
Cereals	3.1	30.6
Legumes	1.0	7.6
Tubers	3.6	17.5
Truck crops	5.7	14.4
Fruits	6.5	10.2
Wine	2.3	16.1
Industrial crops	7.2	3.6
Total	3.3	100.0

Note: Wheat accounted for about 75 per cent of total cereal production in this period, followed by rice (10%), corn (4%), and several more of lesser importance. Among legumes, green beans were about 80 per cent and lentils 12 per cent, green peas and garbanzos accounting for the rest in approximately equal parts. Tubers include potatoes and sugar beets, the former making up well over 90 per cent of the total. No breakdown is available for truck crops, fruits, and wine. Among industrial crops, sunflower seed was most important (35%), followed by hemp fiber (30%) and seed (9%). Flax seed and fiber together accounted for about 10 per cent, as did tobacco. A miscellany of products made up the balance. All figures are from *Indice de producción.*

cated in Table 13.[9] Some of these differences must reflect different definitions of the concepts of "arable," "agricultural," and "irrigated" land. Some sources, for example, count land as irrigated if it is within reach of a canal, even though it may be irrigated only in years of unusually heavy river flow. Other sources count land as irrigated only if it receives water regularly every crop year.

The uncertainties in the land use data make it quite difficult to know if the total amount of land devoted to agriculture changed significantly in the 1950's. Wollman[10] cites some figures indicating that between 1955 and 1962/63 there was little or no change in the total amount of arable land but that the area dedicated to crops may have increased by some 15-20 per cent. However, Wollman notes that this apparent change in land use patterns may in fact reflect no more than changes in classifications.[11]

A Chilean government publication indicates an increase of some 20 per cent between 1949/51 and 1959/61 in the area devoted to those annual crops

[9] A study by ODEPA, completed after the above was written, has cleared up much of the confusion about the amounts of land fit for various uses in Chile. However, the principal question addressed in the text – changes in the amount of irrigated and other land in agriculture in the 1950s – is not clarified by the ODEPA study.

[10] Nathaniel, Wollman, *The Water Resources of Chile* (Baltimore: Johns Hopkins Press for Resources for the Future, 1968), pp. 35–36.

[11] Ibid., p. 36.

Table 13.   Estimates of the Amount of Agricultural and Irrigated Land in Chile

	Irrigated land	Total land fit for agriculture (Thousand hectares)		
		Arable lands	Pasture	Total
La agricultura chilena 1951-1955[a]	1,363	11,079	19,809	30,888
La agricultura chilena 1956-1960	1,366			
Aerophotogrammetric survey[b]		4,600		
CORFO[b]	1,376			
Eduardo Basso S.[b]	1,260			
Dirección de Riego:[c]		5,400	8,300	
Permanently irrigated	1,100			
Occasionally irrigated	865			
Census 1955	1,098	5,514	16,094	
Census 1965	1,103			
Instituto de Investigaciones de Recursos Naturales:[b]				
Permanently irrigated	1,213			
Occasionally irrigated	700			
1950[d]	1,259			
1955[d]	1,338			
1959[d]	1,388			

[a]Ministerio de Agricultura, Departamento de Economía Agrícola, *La agricultura chilena en el quinquenio 1951-1955* (Santiago, 1959).

[b]Cited in Nathaniel Wollman, *The Water Resources of Chile* (Baltimore: Johns Hopkins Press for Resources for the Future, 1968), pp. 33-34. The aerophotogrammetric survey was a joint undertaking of the Chilean government and the Organization of American States. Basso's study is titled *Inventario de recursos hidrológicos superficiales en Chile* (Santiago: Ministerio de Obras Públicas, May 1963). The Instituto de Investigaciones de Recursos Naturales is a part of the Ministry of Agriculture.

[c]Ministerio de Obras Públicas, Dirección de Riego, *Plan nacional de riego: proposición preliminar* (Santiago, July 1967), p. 2.

[d]Carlos Fonck O., Carlos Ladrix H., and Fernando Valpuesta S., "Capital e inversión agrícola en Chile en el decenio 1950-1959" (thesis for the degree of agricultural engineer; Santiago: Universidad de Chile, 1961), p. 22.

accounting for about 62 per cent of total crop output.[12] However, it is reported that revisions of these data, not available as of this writing, show an area increase of only about 10 per cent for these crops in the 1950's.[13]

A study by Fonck, Ladrix, and Valpuesta indicates an increase of 10 per cent in the amount of irrigated land from 1950 to 1959.[14] The data must be

[12]Oficina de Planificación Nacional y Oficina de Planificación Agrícola, *Antecedentes estadísticos sobre superficie, rendimientos, producción y precios agropecuarios* (mimeo.; Santiago, 1966).

[13]Communication from Dr. Delbert Fitchett, formerly with the Rockefeller Foundation, Santiago. James O. Bray asserts that the area seeded to the principal field crops increased by some 10 to 15 per cent in the 1950's. He gives no source for this statement, however. Bray, "Demand and the Supply of Food in Chile," *Journal of Farm Economics,* 44 (1962), p. 1017.

[14]Carlos Fonck O., Carlos Ladrix H., and Fernando Valpuesta S., "Capital e inversión agrícola en Chile en el decenio 1950-1959" (thesis for the degree of agricultural engineer; Santiago: Universidad de Chile, 1961), p. 22. The figures referred to are in Table 13.

viewed with considerable skepticism, however. The census estimates for 1955 and 1965 are well under those of "Capital e inversión agrícola" and show no change in irrigated acreage over that decade. Moreover, the Irrigation Department (Dirección de Riego) of the Ministry of Public Works, the agency most directly concerned with irrigation development, accepts the census figures. Finally, the method by which "Capital e inversión agrícola" arrived at its estimates, while ingenious, has serious weaknesses. Basically it rests on estimates of additional areas to be irrigated given in petitions of private interests for rights to water and on the new areas to be irrigated under public projects. So far as private expansion is concerned, "Capital e inversión agrícola" provides no evidence that the private interests granted water rights actually undertook the investments that would have been necessary to bring under irrigation the area specified in their petitions. Moreover, Wollman makes the flat statement that "virtually all new irrigation in Chile developed in the last fifty years has been in the form of public projects."[15] There undoubtedly was some increase in irrigated area under public projects; the precise amount, however, is highly uncertain. A study by Villarroel and Horn indicates that in the 1950's public projects brought 155,000 hectares under irrigation, 111,000 of them in 1958 alone.[16] (This compares with an increase of 129,000 hectares estimated by "Capital e inversión agrícola" for both public and private projects.) However, the Villarroel and Horn estimate is based on the design area of public projects, and it is well known that there is considerable slippage between design area and area actually irrigated.[17]

In summary, no very definite statements can be made about changes in the 1950's in the amount of arable land, changes in land use patterns, or changes in the amount of irrigated land. The best guesses seem to be that there was no significant change in total arable area, that the amount of land devoted to crops grew by around 10 per cent, and that irrigated area increased less than this, perhaps by some 5 per cent.

Water

It is estimated that about 50 per cent of total agricultural production in Chile in the 1950's was on irrigated land.[18] Accordingly, a significant increase in irrigated area — assuming the supply of other inputs also were larger — would have an important impact on agricultural production. As just noted, the area under irrigation probably did not increase by more than 5 per cent in

[15]Wollman, *Water Resources of Chile*, p. 52.

[16]René Villarroel B. and Heinrich Horn F., *Rentabilidad de las obras de regadío en explotación construidas por el estado* (Santiago: Ministerio de Obras Públicas, Dirección de Planeamiento, June 1963), p. 25.

[17]Wollman, *Water Resources of Chile*, pp. 53–59, cites a number of such cases. One indicates that of a total design area of 78,082 hectares, only 35,403 hectares were effectively irrigated.

[18]Information provided by ODEPA.

the 1950's. If water use moved proportionately with irrigated land, it would follow that greater water use contributed very little to the growth of production in the 1950's. However, the amount of water used per unit of irrigated area was not fixed. Since there were scarcely any regulatory works on Chilean rivers in the 1950's, the availability of water for irrigation depended upon annual precipitation. If the early 1950's were significantly drier than the later years of the decade, this could have had a significant positive impact on the growth of production, even though the total irrigated area had changed very little. Of course, the 50 per cent of output produced on unirrigated land would have been similarly affected by differences in the amount of precipitation at the beginning and end of the decade.

A review of precipitation records collected at 14 weather stations from Copiapó in the north to Punta Arenas in the south indicates that over most of the country there was somewhat more rainfall in 1949-51 than in 1959-61. Ten of the 14 stations recorded more precipitation in the earlier than in the later years, the differences varying from 256 per cent in Copiapó to 6 per cent in Talca.[19] Three stations showed more precipitation in 1959-61 than in the three earlier years, the differences varying from 13 per cent in Aysén to 1 per cent in Puerto Montt. At Los Ángeles average precipitation was the same in both three-year periods. On the average, the 14 stations recorded 10 per cent more precipitation in 1949-51 than in 1959-61.[20] At the 10 stations from Copiapó through Los Ángeles, which encompass substantially all the irrigated area of Chile, precipitation in 1949-51 averaged 16 per cent above the amount recorded in 1959-61.[21]

Thus the early years of the 1950's in Chile were somewhat wetter than the later years. This may have depressed the growth of agricultural production over the decade, although the effect probably was not marked since the rainfall differences were not pronounced. In any case, there clearly is no reason to believe that the growth of production in the 1950's owed anything to greater availability of water.[22]

Labor

Table 14 contains data concerning the number of persons engaged in agriculture in 1952 and 1960. The data show an increase of some 2 per cent

[19]However, recorded annual precipitation averaged only 21 mm. in Copiapó in 1949-51.

[20]This is a weighted average of the relative precipitation at each station in the two periods, the weights being average annual precipitation recorded at each station in 1949-51.

[21]All statements about precipitation in 1949-51 and 1959-61 are based on data taken from *Pluviometria de Chile,* fasc. II (mimeo.; Santiago: Oficina Meteorológica de Chile, 1965).

[22]This statement would not hold if improvements in the efficiency of water use more than offset the decrease in water availability. There is no evidence, however, that water use efficiency increased over the decade.

between these two dates. However, the difference is probably well within the range of error in the two figures, indicating that the number of persons occupied in agriculture was virtually unchanged over the decade. The fact that the percentage of cropland devoted to labor-intensive crops increased over the decade would have tended to increase the average number of days worked per man per year. Offsetting this, however, was a rapid increase in farm mechanization. There is no information on the net outcome of these opposing tendencies; hence we assume that the total number of man-days worked did not change significantly over the decade. It was shown in the last chapter that the rural illiteracy rate declined somewhat, a suggestion that the average quality of the labor force may have slightly increased. On the whole, however, it appears that the total input of labor services, taking account of both the number and quality of persons employed, showed little if any increase.

Table 14.   Economically Active Population of Chile, Total and in Agriculture, 1952 and 1960

		(Thousand persons)
	1952	1960
Total active population	2,155	2,389
Active in agriculture	648	662

Source: Comité Interamericano del Desarrollo Agrícolo (CIDA), *Chile: Tenencia de la tierra y desarrollo socio-económico del sector agrícolo* (Santiago, 1966), Table XI-7, p. 151. The underlying source is the Census and Statistics Department of Chile.

Note: Economically active population is defined as all males and females over the age of 12 who were working or looking for work on the census date.

Fixed Capital

Table 15 presents information relating to the growth of capital stock in agriculture. Most of the data in the table are from "Capital e inversión agrícola" and are admittedly quite rough. They are based on fragments of information plus detailed analysis of changes in capital stock on a sample of 440 farms. The relations between capital items in this sample were used in many cases to piece together the fragments available for the country as a whole to arrive at nationwide totals. Annual net investment in each category of the stock was estimated by subtracting from annual gross investment an estimate of the amount needed to replace capital used up each year. These estimates of replacement were based on the assumption that the value of each category of capital declined by equal annual amounts over its lifetime. The stock each year was then found by adding net investment that year to the stock of the previous year.

In addition to the uncertainties surrounding the basic expenditure data, there was very little to go on in estimating either the average lifetimes of the various capital items or the pattern of decline in their value over time. Moreover, given the rampant inflation in Chile over this entire period, the deflation of the various expenditure series presented major difficulties.[23]

Apart from these general problems, there are specific questions about certain categories of capital. The table indicates that the investment in irrigation works rose by some 30 per cent between 1950 and 1959. Yet, as noted earlier, there is no convincing evidence that the amount of irrigated land increased by any such amount in that period. While there is no need for a one-to-one relation between the expansion of investment in irrigation works and in irrigated area — some of the investment may be designed to increase the security with which a given area receives water — it seems unlikely that such a large investment increase would not involve a rather substantial increase in area irrigated. The estimates of public investment in irrigation are based on data from the Department of Irrigation and are probably reasonably accurate. As noted earlier, however, "Capital e inversión agrícola" provides no evidence that private interests actually undertook the irrigation investments attributed to them.

"Capital e inversión agrícola" estimates that the stock of machinery and equipment, valued in escudos of 1959, increased by 82 per cent between 1950 and 1959. *El uso de maquinaria*[24] estimates that this stock, valued in dollars of 1962, increased by 122 per cent in this period. There are significant differences between the two sources also with respect to the total values of the stock of machinery and equipment. It is impossible to reconcile these differences with the information available to this writer. The escudo of 1959 was worth about $.95 at the then prevailing rate of exchange. If the escudo estimates of machinery and equipment for 1950 and 1959 are converted to dollars at that rate of exchange and then expressed in 1962 dollars by an appropriate dollar price index for machinery, they still are substantially greater in both years than the estimates of *El uso de maquinaria*. Even if the official exchange rate of 1959 greatly overstated the dollar value of the escudo, important differences remain between the two sets of estimates. Fortunately, both sets are in agreement that there was a very large increase in the stock of agricultural machinery and equipment between 1950 and 1959. For present purposes this is more important than agreement concerning totals.

The figures for the increase in the number of tractors between 1948 and 1960 give additional support to the impression that the stock of farm machinery increased substantially in the 1950's. Tractors accounted for 22 per cent of the total value of the stock of farm machinery in 1950 and for 44

[23] Four price series were used as deflators: a construction index, an index for barbed wire, one for machinery, and the wholesale price index.

[24] See Ch. 2, n. 4.

Table 15. Stock of Capital in Agriculture

	Million 1959 E°			Per cent increases		
				Average annual rates		Total, period
	1950	1955	1959	1950–55	1955–59	1950–59
Irrigation works	141.9	161.5	185.4	2.5	3.5	30.6
Buildings, walls, and other structures	337.1	393.6	439.5	3.1	2.3	30.4
Other land improvements[a]	85.9	90.1	97.7	0.9	1.6	11.4
Planted forests	61.3	82.7	87.2	6.1	1.3	42.3
Fruit plantations, vineyards, and cultivated pasture	90.9	112.3	127.4	4.9	2.5	40.1
Total plantations, etc.	152.2	198.0	214.6	5.4	2.0	41.0
Livestock	425.7	415.3	410.1	-0.5	-0.4	-3.7
Machinery and equipment	94.1	163.3	170.9	11.7	4.7	81.6
Total fixed capital	1,236.9	1,418.8	1,518.2	2.8	1.7	22.7
Total less planted forests	1,175.6	1,336.1	1,431.0	2.5	1.7	21.7
Total less planted forests and livestock	749.9	920.8	1,020.9	4.2	2.6	36.1

Number of tractors	163	73	24	226
(Per cent increases)				
Index of machinery and equipment (1951=100)		166		

Sources:

Data in 1959 escudos: "Capital e inversión agrícola," p. 88.

Machinery in 1962 dollars: *El uso de maquinaria* (See Ch. 2, n. 4), p. 2.

Number of tractors: *El uso de maquinaria.* This source provides no estimate of the number of tractors in 1960, but it does present one for 1965. An estimate for 1960 was derived as follows: the ratio of the change in number of tractors 1955–65 to the number imported (also from *El uso de maquinaria*) in this period was applied to the number imported in 1956–60. The result was an estimate of the change in the stock of tractors between 1955 and 1960. This procedure probably overstates somewhat the increase from 1955 to 1960 because imports in 1961–65 spurted 40 per cent above the level of 1956–60. Hence tractors for replacement were probably a higher proportion of tractor imports in 1956–60 than in 1961–65. (All tractors used in Chile are imported.) Machinery index: Data were obtained from *La agricultura chilena . . . 1951–55* on the numbers of most kinds of imported farm machinery in use in 1951 and 1955. The percentage increase in the number of each type between these two years was weighted by the proportionate value of each type in the stock for 1951 to form an index of the stock of imported farm machinery (roughly 90 per cent of all farm machinery). This was combined with the increase in the stock of domestically produced machinery (also from *La agricultura chilena . . . 1951–55*) to form an index of the stock of all machinery.

aLand clearing and leveling, and improvement of interior roads.

per cent in 1959.[25] The index of the stock of all farm machinery also moves roughly in step with the estimates presented in "Capital e inversión agrícola." The index shows that the stock of all farm machinery increased at an average annual rate of 13.5 per cent between 1951 and 1955, whereas "Capital e inversión agrícola" shows an annual increase of 11.7 per cent between 1950 and 1955.

The only major component of capital that declined in the 1950's was livestock. The estimates for this category are based on census data plus information obtained from the Ministry of Agriculture on numbers and prices of animals existing each year. The number of cattle declined by 4.5 per cent between 1950 and 1959, the number of sheep by 9.2 per cent, and the number of horses, mules, and donkeys by 7.5 per cent. Pigs, goats, and chickens all increased rather substantially, particularly the last, which rose by 50 per cent in the 9-year period.[26]

After all the qualifications about data are made, it still appears justified to conclude that the total capital stock in Chilean agriculture increased by some 20 to 25 per cent from 1950 to 1959. All the major components of the stock, except animals, increased more than this, machinery and equipment exhibiting a particularly sharp rise.

Fertilizers and Pesticides

Table 16 presents the available data on fertilizer consumption in Chile in the 1950's. As indicated in the note to the table, the data for 1951-55 are difficult to evaluate and may be subject to a wider margin of error than the estimates for 1957-60. They probably reflect adequately the overall trends, however. Total consumption rose by close to 100 per cent over the decade. Practically all the increase occurred in the first couple of years, however. This was true for all three types of fertilizers. Although consumption of phosphate and potassium fertilizers remained fairly stable over the balance of the decade at the higher levels attained in the first few years, consumption of nitrates declined rather sharply in the second half.

The proportions in which the various fertilizers were consumed at the beginning and end of the decade were as follows:

Per cent of total consumption[27]

	1951-52	*1959-60*
Nitrates	28.0	21.9
Phosphates	59.4	67.0
Potassics	12.6	11.1

[25] *El uso de maquinaria,* p. 2. (See Ch. 2, n. 4.)
[26] "Capital e inversión agrícola," p. 69.
[27] Derived from Table 16.

Table 16. Fertilizer Consumption in the 1950's

(Thousand metric tons of nutritive content)

	Nitrates	Phosphates	Potassics	Total
1951	8.1	17.3	3.9	29.3
1952	11.8	24.9	5.1	41.8
1953	14.9	36.2	7.4	58.5
1954	15.4	31.0	12.1	58.5
1955	15.2	36.4	5.6	57.2
1957	11.6	34.5	7.7	53.7
1958	11.9	40.3	7.8	60.0
1959	11.4	35.2	5.8	52.4
1960	12.9	39.2	6.6	58.7

Sources:
  1951–55: *La agricultura chilena . . . 1951–1955.*
  1957–60: *El uso de fertilizantes.* (See Ch. 2, n. 53.)

Note: *El uso de fertilizantes* is based on a thorough examination of the data on fertilizer consumption in Chile from 1957 to 1965. This examination revealed some deficiencies in earlier published consumption data, and it suggests that the figures for 1951–55 are subject to a fairly wide margin of error.

Table 17. Consumption of Pesticides, 1952, 1958, 1961–63

(Thousand kg. or liters of active ingredients)

	1952	1958	1961	1962	1963
Herbicides	104	347	178	171	292
Fungicides	2,141	5,334	7,266	6,700	6,687
Insecticides	a	1,106	587	709	789
Others	n.a.	n.a.	4	24	21
Total	2,245	6,787	8,035	7,604	7,789

Sources:
  1952–*La agricultura chilena . . . 1951–1955.*
  1958–*La agricultura chilena . . . 1956–1960.*
  1961, 1962, and 1963–*El uso de pesticidas,* p. 7. (See Ch. 2, n. 64.)

  n.a. Not available.
  aIncluded in fungicides.

Consumption of pesticides in selected years between 1952 and 1961 is presented in Table 17. The estimates for 1952 are no more than very rough approximations. Hence the very dramatic increase in consumption between 1952 and 1958 – 202 per cent according to the table – may either overstate, or understate the true increase. There is little question, however, that pesticide consumption increased enormously more than agricultural production in this period. Consumption of fungicides continued to increase after 1958, but herbicide and insecticide consumption apparently declined.

GENERAL CONCLUSIONS ON INPUTS, OUTPUT, AND PRODUCTIVITY

The material just presented indicates that there was no significant change in the amount of land devoted to agriculture in the 1950's, nor was there much increase in irrigated area. Land devoted to crop production may have increased by some 10 per cent, however. Water availability made no contribution to the growth in production. In fact, it may have depressed it somewhat. The number of persons engaged in farming was approximately unchanged, although there may have been some small improvement in the quality of the farm work force to the extent that literacy is an index of labor quality. The stock of fixed capital increased by roughly 20 per cent in the 1950's. However, the expansion over the course of the decade was quite uneven, the rate in the first half being substantially higher than that in the second half for most components of the stock. Consumption of both fertilizers and pesticides rose sharply, although in the case of fertilizers all the increase occurred in the first few years of the decade. Consumption of phosphate fertilizers increased much faster than total consumption while nitrates in particular lagged well behind.

Since water availability was no greater, and may have been less, and there were at most but small increases in the quantities of land and labor resources employed in the 1950's, virtually all the growth of total output in that period must have been due to increases in the stock of capital and to the use of substantially greater quantities of fertilizers and pesticides. Expansion of the area in crops also contributed a small amount to the growth of crop production. Management skill is a major input in agriculture, but there is little reason to believe it improved significantly in the 1950's. As has been noted, agricultural extension services in Chile were very weak throughout this period. This, in combination with only a slight improvement in general literacy, suggests that managerial ability remained about the same over the decade.[28]

It is of interest to note that the expansion pattern of total output in the 1950's was similar to that of fixed capital and fertilizer consumption; that is, output growth in the first half of the decade was more rapid than in the second half.[29] This pattern is quite evident in crop production. Livestock production behaved differently, but this is not surprising, since the output of animal products is a function of the number of animals and production per animal. As already noted, the number declined somewhat in the 1950's. That

----

[28]Corn seems to have been the only crop for which improved seed varieties were used on a significant scale. The impact of this on the expansion of total crop production was insignificant, however. Economies of scale and removal of some marketing bottlenecks may have contributed something to the growth of output, but we have no information on this score. Compared to the inputs mentioned above, however, their contribution, if any, was probably quite small. The impact of technological improvements not already reflected in the growth of capital, fertilizers, and pesticides also must have been limited. The measured increase in these inputs may not fully allow for improvements in their quality over the decade, but there is little reason to believe these were important.

[29]See Table 11.

production nonetheless increased, indicates that output per animal was rising. This probably owed little to increased use of fertilizers, however. There was, to be sure, an increase in the investment in cultivated pasture, and this probably involved some increase in fertilizer consumption.[30] However, the generally stagnant condition of the livestock industry in the 1950's indicates that incentives for innovations of this sort must have been weak. Consequently, it is likely that most of the increase in fertilizer consumption was related to crop production. Some of the growth in the stock of machinery probably contributed also to the increase in animal productivity, but by far the greater part of the new machinery must have been devoted to crop production.

If it can be said that the increase in crop production was due principally to the increase in nonanimal capital, fertilizers, and pesticides, with an extension of cultivated area of lesser importance, and if the major part of the increase in these inputs contributed to crop production, then it becomes of some interest to speculate about the returns to these inputs.

Crop production increased by 36 per cent between 1949/51 and 1958/60.[31] Let us suppose that 75 per cent of this increase was attributable to greater use of fertilizers, pesticides, machinery, and other fixed capital (not including animals, cultivated pastures, or forests).[32] Let us assume further that one of the alternatives shown below represents the separate contributions of fixed capital and fertilizers and pesticides to the total increase in crop production.

Alternative Percentage Contributions to Increased
Crop Production, 1949/51 to 1958/60

Total increase in production	Contribution of –		
	Nonanimal capital	Fertilizers & pesticides[a]	Other
36.0%	9.0%	18.0%	9.0%
36.0	13.5	13.5	9.0
36.0	18.0	9.0	9.0

[a]Fertilizers accounted for about 90 per cent of the combined value of fertilizer and pesticide consumption.

The increase in the stock of capital associated with crop production was about 35 per cent from 1950 to 1959.[33] Under the assumptions made above, there are, therefore, three alternative estimates of the elasticity of production

[30]The increased investment in cultivated pasture is shown in "Capital e inversión agrícola," pp. 87–88.

[31]*Indice de producción.*

[32]The other 25 per cent of the increase in output would be due to all other factors, such as some increase in the quantity of land, possible small improvements in the quality of the labor force and management, better seed varieties, economies of scale, and other miscellaneous inputs.

[33]Increase in total capital less animals, cultivated pastures, and forests.

with respect to fixed capital: .257, .386, and .514.[34] The elasticity of production multiplied by the average product of capital yields the return to the increase in the stock of capital. "Capital e inversión agrícola" indicates that the average product of capital in crop production throughout the 1950's was about $E^O.25$.[35] If this were the case, then the varying estimates of the incremental return to capital would be as shown below:

Incremental Returns to Fixed Capital Associated with
Crop Production under Varying Assumptions

Elasticity of production	Incremental return (per cent)[a]
.257	6.4
.386	9.7
.514	12.9

[a]Incremental return is the product of elasticity and the average product of capital (= $E^O.25$).

In other words, if the 35 per cent increase in fixed capital accounted for but a 9 per cent rise in crop production, then the return to the increment of capital was 6.4 per cent. If, instead, the addition to capital increased production by 18 per cent, then the return was 12.9 per cent.

This same procedure can be used to derive varying estimates of the return to increased use of fertilizers and pesticides. Consumption of these inputs is estimated very roughly to have increased about 175 per cent between 1950 and 1959.[36] The elasticity of production with respect to these inputs, therefore, given their assumed contributions to the increase in output, was .103, .077, and .051.

[34]The elasticity of production with respect to any input is equal to the percentage increase in production attributable to that input divided by the percentage increase in the amount of the input.

[35]"Capital e inversión agrícola," p. 107, shows that total crop and livestock production in the 10 years 1950–59 was $E^O4.2$ billion (1959 prices). *Indice de producción* indicates that in 1955–57 crop production was 52.3 per cent of total agricultural production. Assuming this proportion for the entire decade, crop production for the 10 year period, valued in 1959 prices, was $E^O2.2$ billion. "Capital e inversión agrícola," p. 87, gives the stock of capital in this period, excluding livestock, cultivated pastures, and forest plantations, as $E^O8.8$ billion (prices of 1959). Hence the average product of capital in crop production is estimated at $E^O.25$. This is probably an understatement since some of the capital included is surely directly related to livestock production. The amount cannot be estimated, however.

[36]On the basis of Table 16, fertilizer consumption is assumed to have been 20,000 metric tons in 1950. On the basis of Table 17, pesticide consumption is assumed to have been 1,800 and 7,500 metric tons in 1950 and 1959, respectively. The percentage increases in fertilizer and pesticide consumption were weighted by rough estimates of the relative value of consumption of each type of product in 1959 to derive the combined percentage increase in consumption of the two products from 1950 to 1959. Prices used to derive values of consumption in 1959 were taken from *El uso de fertilizantes* and *El uso de pesticidas*.

The average product of fertilizers and pesticides between 1950 and 1959 can only be calculated with a large margin of error. It probably was between $E^{o}35$ and $E^{o}45$ (prices of 1959).[37] In this case, the return to the increment of fertilizers and pesticides consumed was as shown below:

Incremental Returns to Fertilizers and Pesticides in
Crop Production under Varying Assumptions

Elasticity of production	Incremental returns ($1959\ E^{o}$)	
	(1)	(2)
.103	3.6	4.6
.077	2.7	3.5
.051	1.8	2.3

(1) Returns in this column correspond to average products of $E^{o}35$.
(2) Returns in this column correspond to average products of $E^{o}45$.

In other words, if the elasticity of crop production with respect to fertilizers and pesticides was 0.103 and the average product of these inputs was $E^{o}35$, then one additional escudo spent on them yielded $E^{o}3.6$ in increased crop production. The interpretation of the other estimates is comparable.

The wide margins of error in these estimates of returns to fixed capital, fertilizers, and pesticides have been repeatedly emphasized. However, the estimates are more likely to be understated than overstated. The assumption that these inputs accounted for only 75 per cent of the increase in output from 1950 to 1959 is probably conservative. Management is the only major input not taken into account in the analysis. As noted earlier, there is no evidence that the quality of management increased very much over the

[37]On the basis of Table 16, it is assumed that fertilizer consumption in 1950 and 1956 was 20,000 and 55,000 metric tons, respectively. Total consumption in the period 1950–59, therefore, was about 485,000 metric tons. *El uso de fertilizantes*, p. 44, indicates that the weighted average price of these fertilizers in 1959 was approximately $E^{o}110$ per ton. Total fertilizer consumption over the decade, therefore, is estimated at $E^{o}53.4$ million in prices of 1959. On the basis of Table 17, interpolating for years for which actual data are lacking, total pesticide consumption in 1950–59 was estimated at 43,000 metric tons. *El uso de pesticidas* indicates that almost 90 per cent of the pesticides used in Chile was "azufre ventilado y sublimado" and that the price of this pesticide in 1963 was about $E^{o}160$ per ton (p. 11 for amount consumed, p. 40 for price). Although prices for 1959 are not available, *El uso de pesticidas* (p. 42) indicates that between 1961 and 1963 prices about doubled. It is arbitrarily assumed here that the 1959 price was $E^{o}100$ per ton. Hence, the estimate of total pesticide consumption in the 1950's is $E^{o}4.3$ million. This obviously is a very rough estimate since both price and quantity of consumption data are quite fragmentary. However, even on the most generous assumption, pesticides would have accounted for less than 10 per cent of combined fertilizer-pesticide consumption. Consequently, even a substantial error in the estimate for pesticides would have little effect on the estimate of the combined consumption of these two inputs. This figure is $E^{o}57.7$ million for the decade as a whole. Since total crop production in the period is estimated at $E^{o}2.2$ billion (see n. 35 above), the average product of fertilizers and pesticides is estimated at $E^{o}38$.

decade. While the area in crops probably increased by some 10 per cent, this could not have contributed very much to the increase in crop production.[38]

The estimates of the return to fixed capital are possibly understated because of the unavoidable inclusion of some kinds of capital not related to crop production, resulting in an underestimation of the average product of capital in crop production. If this improperly included capital were as much as 10 per cent of the total used in calculating the average return to capital, and if the contribution of capital, fertilizers, and pesticides to the increase in crop production were 90 per cent instead of 75 per cent, then the alternative returns to capital would be 8.5 per cent, 12.7 per cent, and 17.0 per cent. The returns to fertilizers and pesticides corresponding to these returns to fixed capital would have been $E^O4.3$, $E^O3.3$, and $E^O2.2$ if the average product of fertilizers and pesticides were E^O35, and $E^O5.5$, $E^O4.2$, and $E^O2.8$ if the average product were E^O45.

Despite all the reservations about the quality of the data, these various sets of estimates appear reasonable. We know of no others on a countrywide basis with which to compare them, but as a rule of thumb the return to fixed agricultural capital in Chile is frequently taken to be about 10 per cent. More interesting, and of much greater importance to this study, is the fact that the estimated rates of return are very similar to those obtained from the detailed analysis of farms in O'Higgins Province, reported in Chapter 4. For 177 nonfruit farms (but including livestock) it was found that the marginal return to machinery and structures was 13 to 14 per cent. While this is in the upper part of the range of returns estimated above, the differences are not great. Moreover, O'Higgins Province is one of the richest agricultural areas in the country. It would not be surprising to find returns to capital there somewhat above the national average. The return to fertilizers and pesticides on the 177 farms was $E^O4.7$; i.e., an additional escudo spent on fertilizers and pesticides yielded an increase of $E^O4.7$ in output. On 88 farms specializing in annual crops, the return to machinery was 15 per cent while that of fertilizers and pesticides was $E^O3.96$.

Thus the returns to fixed capital, fertilizers, and pesticides estimated for the country as a whole find considerable support from the analysis of farms in O'Higgins Province, and vice versa. This permits us to assert with some confidence that the foregoing sketch of the performance of agricultural output, input, and productivity is reasonably correct. To summarize the salient features of that sketch, total agricultural production increased at an average annual rate of only some 2.5 per cent in the 1950's. Production of animal products grew by merely 1.6 per cent annually, but for crop production the rate was about 3.3 per cent. The quantity and quality of land devoted to

---

[38]The data for the sample of farms analyzed in Chapters 4 and 5 indicate that the elasticity of crop production with respect to land was 0.20 to 0.25. In this case, a 10 per cent increase in area devoted to crops would raise production by but 2 to 2.5 per cent, given the quantities of other inputs employed.

agriculture did not change significantly in the decade, although land in crops may have increased by some 10 per cent. Even so, land made little contribution to the increase in total production. Annual precipitation at the end of the decade was somewhat less than at the beginning, indicating that water available to both irrigated and unirrigated areas was less at the end of the decade than at the beginning. The number of persons engaged in agriculture scarcely changed. There may have been some small improvement in labor quality, judging from the literacy rate, but this is unlikely to have been very important. In general, it appears that labor made little if any contribution to the increase in production.[39] The stock of fixed capital increased by roughly 20 per cent. The stock of animals, however, actually declined, while capital devoted to crop production increased by about 35 per cent. The stock of farm machinery rose very sharply, the estimates varying from 82 to 122 per cent. Consumption of fertilizers and pesticides increased more rapidly than that of any other input, rising roughly 175 per cent over the course of the decade.[40]

There is little doubt that most of the increase in crop production was due to the employment of greater amounts of fixed capital, fertilizers, and pesticides. The return to the increment of fixed capital appears to have been some 10-15 per cent, while that to fertilizers and pesticides was very large — probably not less than $E^{o}3$ to $E^{o}4$.

Thus, while the overall performance of Chilean agriculture was weak in the 1950's, the sector was by no means stagnant, particularly that part devoted to crop production. The input mix was shifting steadily, and in some cases swiftly, toward the relatively greater employment of such "modern" productive factors as machinery, fertilizers, and pesticides. The "traditional" factors of land and labor correspondingly declined in relative importance, while output per hectare rose perhaps some 20 per cent and output per man increased roughly in proportion to the rise in production.

[39]This implies no slur on Chilean farm labor. It is merely an inevitable consequence of the fact that neither the quantity nor the quality of labor changed much in the 1950's. The analysis of farms in O'Higgins Province indicates that the marginal contribution of labor to output was larger than that of any other single factor of production. Hence, had the quantity of labor employed increased in the 1950's, the impact upon production would have been significant.

[40]It is difficult to say whether given percentage increases in the use of inputs are large or small: much depends upon the base from which one starts. However, the rate of mechanization in Chile compares favorably with that in Mexico, a country generally considered to have achieved considerable success in modernizing its agriculture. Between 1950 and 1960 the number of tractors on Mexican farms increased at an average annual rate of 9.0 per cent. In Chile between 1948 and 1960 the rate was 10.3 per cent. In 1960 Mexico had 1 tractor for every 436 hectares of cropland, while in Chile that year the ratio was 1 tractor for every 315 hectares of cropland. Fertilizer consumption increased much faster in Mexico in the 1950's than in Chile. Nonetheless, in 1960, Chile consumed about 59 pounds of fertilizer per hectare of cropland while Mexico consumed only 40 pounds. (Data for Mexico from the agricultural censuses for 1950 and 1960 and from Reed Hertford, *Principal Historical and Economic Issues in Mexican Agricultural Development* [Mexico, D.F.: U.S. Dept. of Agriculture and Mexican National Institute of Agricultural Research, Nov. 1967].)

Patterns of Resource Use and Potential for Increased Production

The preceding chapter dealt with the overall performance of the Chilean agricultural economy in the 1950's, focusing particularly on the growth of production and attempting to identify and measure the rates of return to the inputs primarily responsible for it. The analysis could not be pushed very far, both because of serious limitations in the data and because the quantities employed of some major inputs, notably land and labor, did not increase appreciably over the decade. This made it impossible to estimate the marginal returns to these resources, given the measurement technique used. Since the kinds and amounts of inputs employed are importantly influenced by their rates of return, our inability to estimate these for land and labor inevitably left the analysis of the pattern of resource use incomplete. Yet such an analysis is essential for a satisfactory assessment of the performance of agriculture.

This chapter and that which follows attempt to maneuver around the obstacles imposed by deficiencies in the overall statistics by working with data for a sample of farms located in the rich Central Valley of Chile. Naturally this route leads to results of more limited generality than those obtainable from data covering the entire agricultural economy. Since the latter are not available, however, the use of sample data is the only alternative open if any progress is to be made at all. As we shall see, the data used, although drawn from a sample of only about 200 farms, nonetheless permit some rather general statements to be made about the performance of Chilean agriculture in the 1950's.

SOURCES OF DATA AND AREA STUDIED

Around mid-1959 the Department of Agricultural Economics, Ministry of Agriculture, collected data for the crop year 1958–59 from a sample of farms in O'Higgins Province, one of the principal agricultural areas of the country.[1] These data were examined by the author, and it appeared that they were suitable for the proposed investigation. Conversations with persons who had participated in the collection of the data and with others who worked with them strengthened this impression.

O'Higgins Province includes 3.3 per cent of the total arable land of the country. In 1955 it had 4.4 per cent of the country's area under crops, 7.9 per cent of that in fruit farms and vineyards and 6.5 per cent in cultivated pastures. The province's share of national production of three principal commodities is shown in Table 18.

Table 18. Production of Three Agricultural Commodities in O'Higgins Province and Share of National Production, 1955 and 1964-65

Commodity	1955		1964-65	
	Quantity of output	Share of nation (percentage)	Quantity of output	Share of nation (percentage)
Wheat (tons)	35,390	3.4	61,636	5.5
Potatoes (tons)	52,521	8.8	75,924	10.8
Wine (thousand liters)	20,303	6.7	44,603	13.3

Sources: Agricultural censuses of 1955 and 1964-65.

In the two areas of the province from which our data were taken – the Central Plain and the valley of the Cachapoal River – 33 different soil types have been distinguished according to physical characteristics.[2] However, it is possible to reduce the number of soil classes using criteria of fertility and permeability.[3] On this basis, the Department of Agricultural Economics divided the region surveyed into seven sections and came to the conclusion that 90 per cent of the land within the region could be described as being flat or very gently sloped, with soils of average depth, and with few or no limitations concerning types of crops that can be grown. Maintenance of the physical conditions of these soils requires simple management practices, such as adequate protection against erosion.[4] In general, therefore, the soils in the region surveyed are of good quality and differences with respect to permeability and fertility are not marked. Nonetheless, soil quality differences both

---

[1] The crop year runs from May 1 to April 30.

[2] Ministry of Agriculture, "Reconocimiento de suelos de la Provincia de O'Higgins," 1957.

[3] This statement and those that follow concerning soil types and climate in the region surveyed are taken from Ministerio de Agricultura, Departamento de Economía Agronómica, "Organización e ingresos de predios especializados en cultivos anuales" (Santiago, 1962).

[4] Ibid., p. 11.

within and between farms do exist, and an effort has been made in this study to eliminate the productivity effects of these differences.

The climate in the region is marked by relatively light and seasonally uneven rainfall, with moderate temperatures moving within a fairly narrow range. In the Central Plain average annual rainfall is 18 inches (460 mm.). Of this, 30 per cent occurs in the fall, 50 per cent in the winter, and 20 per cent in the spring. On the average there are 7 dry months. Because of the scarce amount and the seasonal variations in rainfall, all crop production in the province is on irrigated land. Regional variations in temperatures within the Central Plain are slight. Seasonal variations, of course, are greater, but still moderate. This can be seen in Table 19.

Table 19. Temperatures in Rancagua and Rengo

			(Degrees Fahrenheit)
	Annual average	January average	July average
Rancagua	57.4	70.7	46.6
Rengo	57.4	70.7	46.4

Source: "Organización e ingresos de predios especializados en cultivos anuales." Rancagua and Rengo are the principal urban centers in the province.

The data collected by the Department of Agricultural Economics were taken from 307 irrigated farms lying in the Central Plain extending from the northern to the southern boundaries of O'Higgins Province, and from the level part of the valley of the Cachapoal River. In general, only level, irrigated areas were considered. The population from which the sample was drawn was defined as all farms in the indicated area with from 1 to 500 hectares of irrigated land. Two random samplings were made, one of 104 farms between 1 and 30 hectares and the other of 203 farms between 10 and 500 hectares. The farms in the first group were located in four districts where small farms were the rule. The farms in the second group were located throughout the area described above, including the four districts from which the first sample was drawn.[5]

Of the 307 farms initially sampled, 97 were excluded in undertaking the present study. The most common reason for exclusion was very small annual production (less than E°200[6] was taken as the guide). It was presumed that farms this small were only part-time activities and that their inclusion would

[5] A description of the two surveys and a summary of the results are given in "Organización e ingresos de predios especializados en cultivos anuales." Presumably the farms in the first sample were excluded from the universe from which the second sample was drawn, although this is not stated. In no case does the same farm appear in both groups.

[6] About $190 at that time.

distort the results. Thirty-nine farms were excluded because they specialized in the production of fruits or wine grapes.[7] The production functions for these farms were markedly different from those for other farms. Moreover, it could not be determined from the data how the investment in trees and vines was valued, or even whether these investments were included at all. The 210 farms remaining after these exclusions had a total irrigated area of 22,051 hectares, about 16 per cent of the total irrigated area of the province.[8]

The data were collected directly on the farm through personal interviews, which lasted from as little as one hour to as much as all day. The interviewers were put through a training course before undertaking the survey. Data were collected on the volume and value of output of each productive activity. Output consumed by the farm family was counted as well as output sold. Own-account production by the resident labor force was not included, nor were the land, labor, and other inputs devoted to this purpose.

An important question concerns the extent to which the sample of farms is representative of the population of farms in O'Higgins Province and in the country as a whole. Since the sample includes only farms which obtained 100 per cent, or virtually 100 per cent, of their production from irrigated land, it obviously is not representative of nonirrigated farms. However, most production in O'Higgins Province, and, as noted earlier, about 50 per cent of national production was on irrigated land. Insofar as the sample is representative of irrigated agriculture generally, therefore, it is representative of a major part of all Chilean agriculture.

As noted earlier, the original sample was randomly selected. Our decision to eliminate certain farms changed the randomness somewhat, most particularly in that farms specializing in fruits and wine were eliminated. However, the number of farms remaining in the sample, the fact that they accounted for about 16 per cent of the total irrigated area in O'Higgins, and the fact that they were drawn from all important producing areas in the province permits considerable confidence that the sample is reasonably representative of the population of nonfruit farms in the province.

With respect to the size distribution of farms, the sample is not fully representative of the situation either in O'Higgins or in the country as a whole. This is primarily because the sample deliberately excluded farms of less than 1 irrigated hectare. Unfortunately there is no information on the distribution of farms in O'Higgins or in the country classified by the number of irrigated hectares; however there are undoubtedly a substantial number of

[7]Farms were considered to be specializing in fruits or wine grapes if these products accounted for more than 50 per cent of total production.

[8]According to the 1955 agricultural census the total irrigated area of O'Higgins Province was 141,000 hectares. (Ministerio de Agricultura, Departamento de Economía Agrícola, *La agricultura chilena en el quinquenio 1951-1955* (Santiago, 1959), table facing p. 24.)

farms with less than 1 hectare.[9] The size distribution of farms in our sample is shown in Table 20. Small farms — those of from 1 to 10 hectares — may well be underrepresented, although in the absence of the necessary data this conjecture cannot be checked.[10]

Table 20.   Size Distribution of 210 Farms in O'Higgins Province

| Size class (hectares) | Per cent in each size class classified by — | |
	Irrigated area	Total area
1-5	6.7	5.2
5.1-10	10.9	8.6
10.1-20	11.4	12.3
20.1-50	26.2	23.8
50.1-100	11.4	8.1
100.1-200	16.2	13.8
200.1-500	17.1	18.1
500 +	0.0	10.0

Source: Ministry of Agriculture, sample survey of farmers in O'Higgins Province, 1959. Hereinafter called Survey 1959.

There clearly are some respects in which the sample of farms is not fully representative of all irrigated farms in O'Higgins Province. That they cannot fully represent all irrigated farms in the country is even more clear. Nonetheless, the sample is sufficiently large in number and represents a sufficiently large proportion of irrigated area and production in O'Higgins to support the view that generalizations can be drawn from it. Naturally these generalizations must be hedged, and where possible their consistency with other independent evidence must be tested. Given these precautions we believe the sample data can be used to reach some interesting general conclusions about the performance of agriculture in O'Higgins and in the great Central Valley of which it is an important part.

The sample farms were studied both as a whole and by group according to product specialization. There are three such groups: Group 1 — livestock and dairy products; Group 2 — annual crops; and Group 3 — all farms not falling

[9]The 1955 census showed that 15 per cent of all the farms in the country were less than 1 hectare. While this was based on total land area of the farms — including both nonirrigated and irrigated land — many of these farms no doubt were irrigated. (*La agricultura chilena . . . 1956-1960*, p. 26.)

[10]Comparison of the size distribution of the sample with the available published data on size distribution would exaggerate the underrepresentation of small farms in the sample. This is because the available data include both nonirrigated and irrigated land, and the proportion of irrigated to nonirrigated land is considerably higher on small than on large farms. Thus if farms are classified by total area the proportion of small farms is less than if they are classified by irrigated area only. The sample data in Table 20 demonstrate this.

into one of the other two specialization groups. To be classified in Group 1 a farm had to obtain more than 50 per cent of its annual production from livestock. To be classified in Group 2 more than 60 per cent of annual production had to be in annual crops. Although these classifications achieve greater homogeneity, most farms in any one group also produce some of the products in which the other group specializes.

CHARACTERISTICS OF COBB-DOUGLAS PRODUCTION FUNCTIONS

Cobb-Douglas production functions were fitted to the several groupings of the sample data. The function has the general form

$$Y = AX_l^{b_1} X_2^{b_2} \ldots X_i^{b_i}$$

where Y is production, the X's are inputs, A is a constant, and the b's are the elasticities of production with respect to the various inputs. The function is linear in the logarithms of the variables so that regression coefficients (the b values) can be calculated by the method of least squares. The b's may be positive or negative but are constant over all ranges of inputs. They may or may not add up to 1. If they do, then the function as a whole shows constant returns to scale. If the sum is less than 1 the function is characterized by diminishing returns to scale, while a sum greater than 1 indicates increasing returns.

While the value of b for any single input need not be less than 1, it usually will be. Hence, in the typical case, an increase in the amount of any input employed, other input quantities remaining the same, will yield a less than proportionate increase in output. For example, if b_i is .2, a 10 per cent increase in the employment of X_i, other inputs constant, will increase output by 2 per cent. An implication of this is that all inputs will usually exhibit diminishing marginal productivity as the amounts employed increase.[11]

Economics does not rule out the possibility of negative marginal productivity, but under the great majority of circumstances positive marginal products would be expected. Hence the inability of Cobb-Douglas production functions to depict conditions of resource use under which marginal products

---

[11] The elasticity of production (b_i) with respect to any input (X_i) can be written

$$b_i = \frac{dY}{Y} \cdot \frac{1}{\frac{dX_i}{X_i}} = \frac{dY}{dX_i} \cdot \frac{1}{\frac{Y}{X_i}}$$

The first form of the expression indicates that if elasticity is less than 1, then a small increase in X_i will yield a less than proportionate increase in Y, meaning that the average product of X_i must fall. The second form of the expression indicates that as the average product of X_i $\left(= \dfrac{Y}{X_i}\right)$ falls, the marginal product $\left(\dfrac{dY}{dX_i}\right)$ must fall proportionately since b is a constant.

pass from positive to negative, or vice versa, is not a serious limitation. That the function yields diminishing marginal returns to increasing quantities of input (other input quantities being the same) of course makes it accord well with ordinary economic reasoning.

The objective in fitting the Cobb-Douglas function to a set of cross-section data for a group of farms is to obtain accurate estimates of the marginal contribution of each input to output. Such estimates make it possible to evaluate the performance of the sample of farms and to compare the productivity of different groups of farms. But the estimates so obtained are subject to a variety of possible errors, some of which may be sufficiently large to render the estimates useless, or even dangerous, if the existence of the error is not recognized. These errors stem basically from lack of knowledge of the precise nature of the input-output relations in the production process and from difficulties in specifying and accurately measuring all the inputs employed. The Cobb–Douglas function, in addition to the characteristics already mentioned, is a single equation model of production. That is, it assumes that there is a unilateral causal relationship running from the inputs employed to the output produced. Under these circumstances the Cobb–Douglas function yields the best estimates of the parameters of the relationships between inputs and outputs.[12] If, however, there is interdependence between the inputs and output, i.e., the quantities of inputs employed are determined to some extent by the quantity of output produced, then the parameter estimates will be biased. It is easy to think of situations which would give rise to this problem. For example, if the weather is substantially better than average, with a correspondingly favorable impact on production, farmers at harvest time will be induced to hire more workers and harvesting equipment than they would in a year of normal weather. The quantities of these inputs employed in this case is in part a function of the quantity of output.[13]

A problem may also arise when two or more of the inputs are correlated with one another. This in fact is a common occurrence. Land and labor, for example, will frequently be positively correlated since farms larger in area will also tend to be larger in number of workers employed. Intercorrelation between inputs may be a problem because it tends to obscure the separate contribution of the affected inputs to output. Statistically this shows up in the form of high variances in the estimated regression coefficients of the inputs, meaning that little confidence can be placed in the accuracy of the estimates.

[12] Earl O. Heady and John L. Dillon, *Agricultural Production Functions* (Ames, Iowa: Iowa State University Press, 1961), pp. 137–38.

[13] Hoch has suggested a method for dealing with the problem of simultaneous equation bias. However, it depends upon combining time series data with cross-sectional data, and we have only the latter. (Irving Hoch, "Estimation of Production Function Parameters Combining Time Series and Cross-Section Data," *Econometrica,* 30 (1962), 34–53.)

This is not an inevitable consequence of intercorrelation in the inputs, however, and it arises in only a few of the functions presented below. In particular, in the function upon which most reliance is placed — that for the entire group of sample farms — the variances of the estimated regression coefficients are quite small.

A problem which affects all estimates of production functions is the inability to include inputs, either because they are not known or because no information is available concerning the quantities used. If these omitted inputs are not correlated with the included inputs, the estimated production elasticities of the latter will not be biased, although the sum of the elasticities will be less than if the omitted inputs were included. However, if the omitted inputs are positively correlated with the included inputs, the estimated elasticities of at least one of the latter will be biased upward. If the correlation is negative the bias will be downward.[14]

Management is the most important input usually left out of production functions, including those presented in this study. The problem is that no one has yet devised a satisfactory index of the "quantity" of management.[15] Nor was it possible to do so here. Management is likely to be correlated with capital, since more able managers are likely to build their assets faster and have access to larger amounts of credit than less able managers. Hence it is likely that failure to include management inputs will impart an upward bias to the estimated regression coefficients of capital inputs.

In most production function studies, including the present one, it is necessary to aggregate inputs which may differ from one another in some important respects. In this study, for instance, labor inputs are measured in terms of man-days worked. Hence a day of unskilled labor counts just as heavily in the production function as a day worked by the most highly skilled manager. It has been shown that ignoring quality differences in inputs is equivalent to the omission of a number of variables together with the inclusion of an additional variable.[16] The omitted variables are those for the different quality inputs, while the additional variable is the conglomerate representing all of them. Omission of inputs will give an upward bias to the regression coefficient of the conglomerate input. The overall bias, however, will depend upon whether the inclusion of the conglomerate reinforces or offsets the omitted variable bias.[17]

The great majority of labor employed in Chilean agriculture in the late 1950's was unskilled. Hence failure to allow for quality differences in labor

[14] Zvi Griliches, "Specification Bias in Estimates of Production Functions," *Journal of Farm Economics*, 39 (1957), 8–20. Also Heady and Dillon, *Agricultural Production Functions*, p. 214.

[15] Heady and Dillon, *Agricultural Production Functions*, p. 224.

[16] Ibid., pp. 215–16.

[17] Ibid., p. 216.

inputs probably does not seriously bias the corresponding regression coefficients. Nonetheless, the probable existence of some bias should be kept in mind.

In general, omission of relevant inputs, some of which are correlated with included inputs, will tend to give an upward bias to the sum of the elasticities computed in Cobb–Douglas production functions. Empirical studies by Hoch and Mundlak suggest that this bias may be important.[18] Both employed the technique of covariance analysis in computing farm production functions and compared the results with those obtained from Cobb–Douglas functions fitted to the same data. Hoch found that the sum of the regression coefficients fell from .954 to .832, although the behavior of coefficients for particular inputs was mixed. The labor coefficient fell sharply from .241 to .057, while that for feed, fertilizers, and pesticides fell from .315 to .288. Those for other current expenses and fixed capital rose modestly.[19] Mundlak found that the sum of the regression coefficients fell from .967 to .878. Those for labor and variable operating expenses dropped most sharply, whereas that for barns and livestock rose.[20]

In principle it is desirable to measure fixed capital inputs by the annual flow of capital services rather than by the stock in existence at any given time. Unfortunately, in this study the data did not permit measurement of the flow of capital services. Instead, fixed capital inputs are represented by the value of the stock in existence at the end of the year surveyed. It has been demonstrated that the difference between capital stocks and flows can be quite important in computing production functions.[21] In a study of 430 Greek farms Yotopoulos calculated Cobb–Douglas production functions using both stocks and flows of fixed capital. He found that the regression coefficients for land, plant, and "live capital" (animals and trees) were not statistically significant in the "stock" function, but that in three different "flow" functions each of them was of increased importance and also significant. Equipment was significant in both types of functions, but of somewhat less importance in the flow types. The most startling change was in live capital, the coefficient of which increased from .05 in the stock function to .26 in the flow functions. "*T*" values rose from slightly less than 2 to over 7.

This discussion leaves no doubt that production function analysis of farm operations provides no magic formula for quantifying the contribution of each input to final output. Hence it cannot be pretended that the technique

---

[18] Hoch, "Estimation of Production Function Parameters"; Yair Mundlak, "Empirical Production Function Free of Management Bias," *Journal of Farm Economics*, 43 (1961), 44–56.

[19] Hoch, "Estimation of Production Function Parameters," p. 45.

[20] Mundlak, "Empirical Production Function," p. 52.

[21] Pan A. Yotopoulos, "From Stock to Flow Capital Inputs for Agricultural Production Functions: A Microanalytic Approach," *Journal of Farm Economics*, 49 (1967), 476–91.

permits precise statements about the performance of farmers or about relative productivity differences between groups of farms. However, if the results of production function analysis are consistent with independent information concerning farm operations, and the analysis of them is leavened with a healthy dose of common sense, useful statements can still be made. It is that spirit which animates the following discussion.

ANALYSIS OF PRODUCTION FUNCTIONS

Nine separate sets of inputs were included in the final calculations (although not all of them appear in each function): two land variables, a labor variable, three operating capital variables, and three fixed capital variables.

Land

X_1 All irrigated land, measured in hectares, excluding that in roads or paths, under houses or other structures. It is directly productive land.

X_2 X_1 adjusted to allow for differences in the quality of land. The adjustment was based on information in the tax rolls for each farm. The land assessments for tax purposes are supposed to be based principally on soil quality.[22] The relation between these assessments for different soils, therefore, provides an index of soil quality differences both within farms and between farms. Land of top quality was given a rating of 1 in our scheme (i.e., for this land $X_1 = X_2$). Land with lesser quality soils was scaled down accordingly (i.e., for this land $X_1 > X_2$ in the proportion in which its soils are less than prime quality). The information necessary for making this adjustment was available for 177 of the 210 farms included in the sample.

X_3 Labor inputs measured in man-days worked per year. No distinction was made between types of labor.

X_4 "All other" operating expenses, i.e., total operating expenses less the value of fertilizers and pesticides (X_6), and seed (X_5). Included are outlays for fuel, machinery repairs, feed, and so on.

X_5 Value of seed used, whether purchased or grown on the farm. In the latter case it was valued at the market price.

X_6 Purchases of fertilizers and pesticides, with fertilizers accounting for from 80 to 90 per cent of the total.

X_7 End-of-year market value of machinery and equipment as estimated by the farmer. Automobiles and trucks primarily for the personal use of the farmer or his family were excluded.

X_8 Average of beginning-of-year and end-of-year value of all livestock as estimated by the farmer.

---

[22] That is, the value of improvements such as pastures, fences, buildings, orchards, etc., are left out of account.

X_9 End-of-year depreciated value of all productive structures – barns, wells, fences, walls, irrigation ditches, and the like. Housing for the owner and his resident labor force, if any, was excluded.

As stated before, no information was available for management inputs. This was true also for water. While every farmer knows what his water rights are, these are usually translated into so many hours per day or days per week in which he has the right to use whatever is flowing through the canal. The amount of water actually used, however, is one of the great unknowns of Chilean agriculture. This is unfortunate, not only because it deprives the present study of useful information, but because without such information returns to water cannot be calculated and policy regarding water use must be made almost blind. Since all the farms in our sample are fully or almost fully irrigated in the sense that all or virtually all their production is from irrigated land, the failure to include water inputs is not so serious as it would be if the sample included both unirrigated and irrigated farms. Water use is probably correlated with the amount of land irrigated. Hence, the failure to include water may impart some upward bias to the estimated regression coefficients for land.

Tables 21 and 22 present Cobb–Douglas production functions, marginal value products, and other information for various groupings of farms. For any input the marginal value product (MVP) was calculated as the product of the input's regression coefficient (elasticity of production) and its average product calculated at the geometric means of output and input.

Table 21.   Cobb–Douglas Production Functions and Other Information for Farms in O'Higgins Province, 1958/59

	Regression coefficients		Marginal value products ($E°$)	
	210 non-fruit farms	177 non-fruit farms adjusted[a]	210 non-fruit farms	177 non-fruit farms adjusted
	(1)	(2)	(3)	(4)
Land (X_1)	.102[c] (.060)		12.18	
Land, adjusted (X_2)		.235[b] (.059)		54.22
Labor (X_3)	.375[b] (.061)	.260[b] (.062)	1.28	0.95
All operating expenses except fertilizers, pesticides, and seeds (X_4)	.345[b] (.039)	.308[b] (.042)	1.60	1.51

Table 21. *Continued*

	Regression coefficients		Marginal value products (E^O)	
	210 non-fruit farms	177 non-fruit farms adjusted[a]	210 non-fruit farms	177 non-fruit farms adjusted
Seeds (X_5)	.086[b] (.027)	.095[b] (.030)	1.58	1.77
Fertilizers and pesticides (X_6)	.067[b] (.017)	.062[b] (.018)	5.50	4.73
Machinery (X_7)	.055[c] (.027)	.049[c] (.028)	0.14	0.13
Livestock (X_8)	.006 (.022)	.012 (.023)	0.02	0.04
Productive structures (X_9)	.016 (.012)	.030[b] (.013)	0.08	0.14
Sum regression coefficients	1.042	1.050		
R^2	.958	.964		
A (intercept of production function)	−.106	.117		
Value of F testing significance of regression	568.2	558.4		

Notes:
Figures in parentheses are standard errors of the regression coefficients.
The marginal value products for machinery, livestock, and productive structures can be interpreted as per cents. For example, the marginal return to machinery investments on the 177 farms was 13 per cent. These are real rates of return despite the chronic inflation in Chile. This is because the prices of both the inputs and outputs refer to a common period of time, the end of the crop year. Only if farmers believed that the future prices of these inputs would change in relation to output prices would the calculated rates of return not be real rates of return.

[a]There were 177 of the 210 farms for which it was possible to adjust the land variable for quality differences.
[b]Regression coefficient significantly different from zero at 1 per cent probability level in a one-tail test.
[c]Regression coefficient significantly different from zero at 5 per cent probability level in a one-tail test.

The F values obtained in testing for the significance of the regression equations (production functions) show them all to be highly significant, i.e., the observed relationships between the outputs and corresponding sets of inputs could not conceivably be attributable to chance factors. The values of R^2 show that in every case the joint influence of the inputs explains 95 per cent or more of the corresponding outputs.

Table 22. Cobb–Douglas Production Functions and Other Information for Farms in O'Higgins Province, Classified by Product Specializations

	Regression coefficients			Marginal value products ($E°$)		
	35 livestock farms	88 annual crop farms	54 unspecialized farms	35 livestock farms	88 annual crop farms	54 unspecialized farms
Land, adjusted (X_2)	.106 (.159)	.231[a] (.081)	.208[b] (.107)	29.15	53.25	42.87
Labor (X_3)	.006 (.179)	.319[a] (.091)	.194[b] (.108)	0.03	1.19	0.61
All operating expenses except seeds, fertilizers, and pesticides (X_4)	.259[a] (.091)	.315[a] (.069)	.386[a] (.087)	1.03	1.83	1.65
Seeds (X_5)	.160[c] (.109)	.108[a] (.040)	−.004 (.052)	3.87	1.87	−0.06
Fertilizers and pesticides (X_6)	.020 (.040)	.069[a] (.028)	.123[a] (.035)	2.36	3.96	11.38
Machinery (X_7)	.084 (.085)	.055[c] (.041)	−.048 (.046)	0.22	0.15	−0.13

Livestock (X_8)	.231[a] (.091)	-.029 (.029)	.139[b] (.062)	0.43	-0.12	0.32
Productive structures (X_9)	.086[b] (.041)	.013 (.022)	.022 (.018)	0.31	0.06	0.12
Sum regression coefficients	.953	1.085	1.020			
R^2	.983	.953	.976			
A (intercept of production function)	.687	-.016	.265			
Value of F testing significance of regression	186.4	199.5	231.6			

Notes:

Figures in parentheses are standard errors of the regression coefficients.

Functions were also calculated using unadjusted land. These differed from those shown above mainly in yielding higher variances in the regression coefficients for land.

[a] Significantly different from zero at 1 per cent probability level in a one-tail test.
[b] Significantly different from zero at 5 per cent probability level in a one-tail test.
[c] Significantly different from zero at 10 per cent probability level in a one-tail test.

The sums of the elasticity coefficients are very close to 1 in all cases. The previous discussion of the several biases likely to affect elasticities estimated from Cobb–Douglas production functions suggests that these sums may be too high. If management inputs were included, for example, the sums of the elasticities for the other inputs would probably be less. If the studies of Hoch and Mundlak can be taken as a guide, the labor coefficients in particular may be overstated.

Elasticity Coefficients in Table 21

The estimated coefficient for land inputs on the 210 farms is significant at the 5 per cent level of probability. After adjustment for quality differences, the estimate for land improves, being significantly different from zero at the 1 per cent level of probability.[23] It is interesting that the adjustment for soil quality increases the value of the regression coefficient. This result makes sense since it indicates that the marginal contribution to output of better quality land is higher than that of lower quality land.[24]

In the function for 177 "adjusted" farms the coefficients for all inputs except livestock are significantly different from zero at the 5 per cent level of probability or less. Most of them are significant at the 1 per cent level. Except in reference to livestock and productive structures this statement is true also for the group of 210 "unadjusted" farms. In both functions the coefficients for labor and the collection of inputs included in X_4 are the two most important inputs; together they account for from about 55 to 70 per cent of total production. The earlier discussion of possible biases suggested that labor coefficients may be subject to some upward bias. Even so, it is almost certainly true that labor was one of the two major contributors to output among the farms sampled.

The coefficients for livestock in both functions are not significantly different from zero. This may reflect some defect in the data – although none are obvious in the original sources – or it may result from the treatment of animal capital as a stock rather than as a flow. In the previously cited study by Yotopoulos[25] it was found that the animal coefficient rose dramatically and its variance was sharply reduced when animal inputs were measured as flows rather than stocks. Nonetheless, it was shown in Chapter 3 that the overall performance of the livestock industry was weak in the 1950's, and it

---

[23] The improvement in this estimate may not be due entirely to the adjustment for soil differences, since the number of farms in the adjusted function is different from the number in the unadjusted function.

[24] The result also suggests that the evaluation of land by the Chilean internal revenue department is at least in the right direction, i.e., more productive land generally is assessed higher than lower quality land. Needless to say, the results do not demonstrate that assessments *exactly* reflect productivity.

[25] See n. 21 above.

may well be that for the general run of farmers the marginal return to this activity was little if any different from zero.

It is to be expected, of course, that the inputs used in agricultural production will make a positive contribution to output. The production functions presented in Table 21 are in general quite consistent with this expectation, since one can say with considerable confidence that almost all the inputs appearing in them actually did contribute positively to output. The only consistent exception, livestock, may in fact have contributed little or nothing, as just noted.

Elasticity Coefficients in Table 22

At first glance this last statement appears inconsistent with the returns to livestock on livestock farms and on unspecialized farms shown in Table 22. The estimated coefficient for livestock on livestock farms is significant at the 1 per cent probability level and the marginal return is 43 per cent. On unspecialized farms the coefficient is significant at the 5 per cent level and the marginal return is 32 per cent. Are these results consistent with the finding that when farms are not distinguished with respect to specialization, returns to livestock are not significantly different from zero? There is always the possibility of errors in one or more of the estimated coefficients. However, the differences between the estimates may have an economic rationale. Virtually all the 177 farms included in the sample had some animals kept for productive purposes (aside from work animals such as horses and mules), beef and milk cows being the most important. On only 35 farms, however, did livestock contribute more than 50 per cent of total production. It is possible that on annual crop farms livestock production was considered a sideline, the farmer's main effort being devoted to production of annual crops. In this case his knowledge of and attention to proper animal management practices could have been considerably less than those of farmers specializing in animal production. In this case, the marginal return to animals on farms especially devoted to that activity could easily be well above that on farms specializing in something else.

This explanation is consistent with the finding for unspecialized farms. While these farms did not devote as much attention to livestock as farms specializing in that activity, their investment in animals was, on the average, 33 per cent greater than that on annual crop farms. Moreover, animals represented 30 per cent of the total fixed investment on unspecialized farms compared with 20 per cent on annual crop farms. Thus it is not implausible to find the return to animals on unspecialized farms higher than on annual crop farms, although not as high as on farms specializing in animal production.

The regression coefficients for land and labor were significantly different from zero on both annual crop and unspecialized farms, but on livestock farms they were not. The latter is a suspicious result, and may possibly be due to the relatively small number of farms in the livestock group. The experience of a mere 35 farms may not be sufficient to reveal the true pattern of relationships in the production function, particularly when 8 inputs were specified in the function. Another possibility with respect to livestock farms is that total land area may be a better measure of land inputs than irrigated area. However, substitution of total area for irrigated area in the production function for these farms did not change the result.

The uniformly high R^2's in Table 22 are consistent with those in Table 21 and are almost certainly overstated. Aside from the omission of management, there is high intercorrelation among the independent variables. This is particularly true of the function for livestock farms, where the correlation coefficients between pairs of independent variables are almost all over 0.9. (This also may in part explain the high variances in the estimates for some inputs.) For annual crop farms and unspecialized farms the intercorrelations are less pronounced, although between land and labor they are about 0.9, while between these two and X_4 (in the case of annual crop farms) and X_{11} (for unspecialized farms) they are 0.80 to 0.85.

Of course, no pretense is made that the results presented in Tables 21 and 22 are precisely accurate in all details. It is believed, however, that the pattern which emerges is sensible and correct in its major features. This belief in part is grounded in the writer's judgment that the underlying data are basically sound. More important, however, is the fact that except for land, livestock, and fertilizers, the marginal returns to the various inputs are generally consistent with what is known about the costs of these various resources. The exceptions do not upset the general conclusion because plausible explanations for them are at hand. In the final analysis, therefore, the case for the plausibility of the results shown in Tables 21 and 22 rests on their consistency with independent information concerning the conditions of Chilean agriculture in the 1950's. The reader is not asked to take this assertion on faith. The argument for it is developed at length in the following section.

Patterns of Resource Use

The production functions provide a useful point of departure for analyzing the performance of the farms in the sample, and by inference, of the population of farms from which the sample was drawn. The focus of the analysis is on the patterns of resource use, for this will reveal whether the farms, as a group, were employing the most economical amounts of each input or whether they could have profitably increased their use of some and decreased that of others. Inputs are employed in the most profitable quantities when their MVP's are just equal to their prices. If the MVP of any input exceeds its price, then increased use of it will add more to production than to cost; hence it will "pay" to use more of the resource. The reverse is true if the MVP is less than the price — it will "pay" to use less of the resource — while equality of marginal products and prices indicates that no gain in production can be achieved through either more or less use of the input. Hence study of the differences between marginal value products and prices of inputs should be useful in explaining the pattern of use of resources in agriculture and in identifying those inputs which promised the largest increases in output from greater employment of them.

To simplify the discussion, attention is focused on the 177 nonfruit farms represented in columns (2) and (4) of Table 21. The principal reason for this is that the production function for these farms gives reliable estimates of input coefficients and represents a wide variety of farming activities. To be sure, some, perhaps all, of the coefficients in this function are biased to some extent for reasons already discussed. This is true of all the functions, however. Needless to say, the conclusions reached do not necessarily apply to any single farm, nor do they apply in full detail to each of the three specialization groups.

As noted above, the analysis is focused on the differences between the marginal value products and the prices of the various inputs included in the production function. In this context the "price" of any input is the cost to the farmer of using one more unit of it. For land, machinery, and other fixed capital items this cost is stated as a percentage rate of return so as to be comparable to the marginal products of the assets, which are also measured in percentages.[26] The relevant cost in these cases is the maximum percentage return to be had from alternative investments open to the farmer, allowance made for risk. These alternatives include those both on and off the farm. The farmer has the option of switching the mix of land and machinery, for example, but he also has the opportunity of investing in nonagricultural sectors of the economy.

Unfortunately, very little is known about rates of return to nonagricultural investments in Chile. However, judging from the rate of growth in manufacturing production and output per man between 1957 and 1961, there must have been very attractive investment opportunities in that sector. Value added in manufacturing, in constant prices, rose by one-third in the 4-year period, while output per man increased by 23 per cent. Rates of expansion in some major industry groups were substantially higher than for manufacturing as a whole.[27] In view of this performance we think it unlikely that real returns in manufacturing could have been below 10 to 15 per cent, but in the absence of more evidence there is no point in pushing the matter. In what

---

[26] In the production function, the marginal product of land is not measured as a percentage return, but it is converted to that basis in the comparisons that follow.

[27] Corporación de Fomento de la Producción, *Geografía económica de Chile* (Santiago, 1966), p. 154.

follows it will be assumed that returns to agricultural investments of less than 10 per cent were below equilibrium levels, while those much above 15 per cent were high.

At first sight it might appear that the relevant cost of fertilizers, seeds, pesticides, and other operating inputs is E^O1. That is, these inputs ought to be purchased up to the point where their marginal products are E^O1, since their marginal contribution to production would then just equal their cost. If the cost of using an additional unit of any of these inputs were simply its price, this reasoning would be correct. However, the greater use of any of these inputs generally involves costs in addition to the cost of the input itself. If the farmer borrows to finance these inputs he must expect to pay interest. Even if he has the good fortune to borrow under inflationary conditions such that the real rate of interest is negative, he must consider the returns to these funds in all their alternative uses, including personal consumption. The use of resources currently to yield a deferred income, even if for only one crop year, always entails an implicit interest charge.

But perhaps the most important factor leading to a divergence between the prices and marginal products of flow inputs is risk. A farmer may know quite definitely that an additional unit of input A will cost E^O1, but he will have only a more or less firmly held expectation that the return will be, say, $E^O1.3$. On the basis of his own experience and that of his neighbors he may believe that this is the most probable outcome. He will know, however, that other outcomes, while less probable, are still well within the range of possibility. Some of these may even be less than E^O1. The decision of any particular farmer to use more of input A will be conditioned by what he thinks he knows of the distribution of probable outcomes, and also by his attitudes toward risk taking. It would not be surprising, however, to find that in general farmers insist on a most probable output response significantly greater than E^O1 before deciding to spend an additional escudo on input A.

In the Chilean case it is probably necessary to make some allowance also for the effects of inflation on the relative prices of outputs and inputs. In an inflationary setting farmers will not employ inputs to the point where an additional expenditure of E^O1 yields a return of E^O1 because the value of the escudo will decline between the time of purchase of inputs and the time output is sold. In these cases, therefore, the farmer will demand an "inflationary premium" in the marginal value products of the inputs. The amount of the premium would depend upon the rate of inflation in input prices and upon the average time lag between the purchase of inputs and the sale of outputs. There is no reliable information on the rate of inflation of input prices in 1958/1959, nor do we know anything definite about the time lag between input purchases and output sales. All that can be said is that some allowance for this factor ought to be made in estimating the real costs of flow inputs such as fertilizers, pesticides, seeds, and fuel.

On the basis of these considerations it was concluded that in the comparison made here the relevant cost of fertilizers, seeds, and other flow inputs is in the neighborhood of $E^O1.50$.

The relevant cost per man-day of labor employed on these farms was the average wage actually paid. This came to $E^O.83$ per day, including payments in kind.

The production function yields marginal returns to land as so many escudos per hectare per year. To compare this return with the cost of capital to the farmer it is necessary to convert it to a percentage basis. This requires information on the price of land like that used by the 177 farms. Such information is very scarce. The best available is based on a study of prices of average quality irrigated land in the region between the Bio-Bio River in the south and the Province of Atacama in the north (O'Higgins lies in the southern part of this region). Both improved and unimproved land were included in the study. On the basis of the study, a price per hectare of E^O460 seems a reasonable approximation for the crop year 1958/1959.[28]

This figure, however, is for average quality land. The land variable included in the production function for the 177 farms was adjusted in such a way as to express each farm's land in units of prime quality land. Hence what is needed is a price for prime quality land in 1958/1959. This price is not available. However, conversation in Chile with economists with long experience in the country and with a special interest in agriculture suggests that prime quality land could easily have been priced at twice the value of average quality land. Hence E^O900 per hectare is taken as the price of land in calculating the rate of return to land investments. Since the marginal product of land on the 177 farms was $E^O54.22$ per year, the return to land priced at E^O900 per hectare was 6 per cent.

Table 23 compares the marginal products of the various inputs with their respective costs. The table indicates that the return to land on nonfruit farms

[28]The study referred to is by Tom Davis, "Estudio sobre el precio real de la tierra agrícola a partir de 1928" (Santiago: Universidad Católica de Chile, Centro de Investigaciones Económicas, July 1958). Davis presents an estimate for the early months of 1958 of E^O450 per hectare. Lagos has extended the Davis estimates to 1962 and presented the results in index form. The change in this index from 1958 to 1959 was used to estimate the price for the latter year. The Lagos index is in Kurt Ullrich B. and Ricardo Lagos E., *Agricultura y Tributación* (Santiago: Universidad de Chile, Instituto de Economía, 1965), p. 80. In his study of the comparative performance of private farms and those sponsored by the Chilean land reform agency, Aldunate estimated the average price of unimproved irrigated land in the Central Valley at $E^O1,500$ per hectare in 1963/64. If one adjusts this figure to 1962 price levels by means of the index of food prices to consumers (a very rough proxy for an index of land prices) and from 1962 to 1959 by means of the Lagos index, the result is E^O570 per hectare. This is close enough to provide some support for our estimate of E^O460. (Paul Aldunate, "A Comparison of Resource Productivity and Efficiency on Private and Government Created Farms in the Central Valley of Chile," M.A. thesis in agricultural economics, Purdue University, June 1965.)

in O'Higgins was almost surely less than the cost of capital.[29] While the return to livestock was even less than that to land, the variance of this estimate is so large that it cannot confidently be stated that the return was in fact less than the cost. This is true also of the return to labor. Although it was above the average wage paid by the 177 farms, the difference was not large. Similarly, while the returns to seeds and operating expenses other than fertilizer were different from the assumed costs, the differences are insignificant. In the case of fertilizers, however, there is little doubt that the marginal return was well above cost.

The pattern of marginal returns and costs in Table 23 points toward a very important conclusion: the supply of land was not the factor limiting the expansion of agricultural output in O'Higgins Province in the late 1950's. *Given the supply of fertilizers, seeds, machinery, and other nonland inputs, it would have been uneconomical for farmers as a group to purchase or rent more irrigated land, since the return to land was already well below the cost of capital.* However, given the amount of irrigated land devoted to nonfruit production, a considerably greater quantity of *all* nonland inputs, except animals, could have been profitably employed. The principal reason for this is the very high return to fertilizers, an indication that it would have paid to use substantially more of this input. Had this occurred, the marginal returns to labor, machinery, seeds, other operating expenses, and productive structures would have been raised above their respective costs. Hence the use of substantially more fertilizers would have justified the use of more of these other inputs also.

The production function can be employed to estimate the increased amounts of fertilizers and other nonland inputs which could have been profitably employed, *with no increase in the amount of irrigated land*, and the increase in nonfruit production which would have resulted.[30] These estimates are shown in Table 24.

---

[29] Applying conventional standards, the differences in Table 23 are considered significant only if the probability of their occurrence was less than 5 per cent. By this criterion only the differences for land and fertilizers were significant.

[30] Table 21 shows that the sum of the regression coefficients in the function for the 177 farms was 1.05, indicating increasing returns to scale. This means that in principle these farmers could have indefinitely increased output by 10.5 per cent for every 10 per cent increase in all inputs *including land*. Even though the return to land would have continued to be less than the opportunity cost of capital had this course been followed, it obviously would have paid if in fact proportionate increases in inputs had yielded more than proportionate increases in output. This was not a real possibility, however. In the first place, an indefinite expansion of all inputs would not have been feasible, irrigated land in particular being in quite inelastic supply for any large group of farmers taken as a whole. Moreover, individual farmers would have been aware of the generally low return to land and the high return to fertilizer. The natural thing in this case would be to try to increase the amount of fertilizer while holding the amount of land constant. The important question, why this did not occur, is treated at length in the discussion that follows.

Table 23. Marginal Value Products (MVP) and Costs of Inputs on 177 Nonfruit Farms in O'Higgins Province, 1958/59

	MVP	Costs	Difference between MVP and costs	Probability of observed difference if true difference were zero
Land (per hectare, adjusted)	6%	10%–15%	4%–9%	1%–less than .1%
Labor (per man-day)	E°0.95	E°0.83	E°0.12	51%
All operating expenses except fertilizers, pesticides, and seeds	E°1.51	E°1.50	E°0.01	96%
Seeds	E°1.77	E°1.50	E°0.27	63%
Fertilizers and pesticides	E°4.73	E°1.50	E°3.23	.5%
Machinery	13%	10%–15%	–	–
Livestock	4%	10%–15%	6%–11%	34%–8%
Productive structures	14%	10%–15%	–	–

Sources: MVP from Table 21. Costs were derived as described in the text. Since the MVP's for machinery and productive structures fall between 10 and 15 per cent, they were assumed to be, in effect, equal to the cost of capital.

Table 24. Actual and More Profitable Patterns of Resource Use and Output on Nonfruit Farms in O'Higgins Province, 1958/59

	Output (E°)			MVP corresponding to more profitable input quantities[a]	Input quantities[b]		
	Actual	More profitable	Col. (2) ÷ (1)		Actual	More profitable	Col. (6) ÷ (5)
	(1)	(2)	(3)	(4)	(5)	(6)	(7)
Land				10%	27.9	27.9	1.00
Labor				0.95	1,756	2,850	1.62
Operating expenses except fertilizers, pesticides, and seeds				1.52	1,315	2,105	1.60
Seeds				1.60	345	615	1.78
Fertilizers and pesticides				1.51	84	420	5.00
Machinery				11%	2,414	4,825	2.00
Livestock				6%[c]	2,157	2,157	1.00
Structures				13%	1,439	2,430	1.69
Output	6,447	10,360	1.61				

[a]In the case of land, machinery, livestock, and structures these figures should be interpreted as the annual rates of return on additional investments in these assets. For labor the MVP shows the increase in the value of production resulting from an additional man-day of work. The MVP's for all other inputs show the contribution to production of the expenditure of one additional escudo.

[b]Land is measured in hectares, labor in man-days, and all other inputs in escudos.

[c]The increase in all other resources needed to raise the return to livestock to 10–15 per cent would be so large as to make the exercise illustrated in this table of doubtful meaning. Hence the table indicates the quantities of inputs consistent with efficient use ignoring the return to livestock.

The estimates of marginal value products corresponding to more profitable quantities of inputs (column 4 of Table 24) were found by successive substitution of inputs in the production function until a set of marginal products were obtained which while not exactly equal to the assumed costs of inputs, are not significantly different from them. (The more profitable quantities of inputs are shown in column 6 of Table 24.) The results shown in the table should be interpreted with caution. The production function measures the average relation of each input to output at the geometric mean levels of use of each input. Assuming that the sample data are representative, one can specify with known probability the range of output levels which would be obtained if the mean quantities of inputs were employed. That is, one can make a statement such as: If a group of farms like those from which the sample was taken were to employ the mean level of each input, then there is X per cent probability that the resulting output would lie between Y and aY. As the probability of the statement rises, the output range increases.

As one moves away from the mean levels of input, the output range corresponding to any given level of probability widens. Looked at differently, the probability level corresponding to any given range of output is less. Thus if another sample of farms were drawn from the same population as our 177 farms and employed the same quantities of each input as the 177 farms, then there is a 95 per cent probability that the resulting mean output would fall between E⁰5,400 and E⁰7,484.[31] This means that if the output interval corresponding to a given level of probability were invariant with respect to the quantities of inputs, then it could be said with 95 per cent confidence that had the 177 farms employed the more profitable quantities of inputs shown in Table 24, the resulting output would have been between E⁰9,313 and E⁰11,407. However, because the inputs shown are in most cases above the mean levels, the confidence level corresponding to an output interval of E⁰9,313 to E⁰11,407 is less than 95 per cent. It is still quite high, however.

No claim is made that the 177 farms would have produced precisely 61 per cent more output if they had employed the more profitable quantities of inputs shown in Table 24. It is asserted, however, that the table presents strong evidence that farms in O'Higgins Province in the late 1950's could have profitably employed a substantially larger quantity of all nonland inputs (except livestock), *including labor.* Since labor inputs are measured in man-days worked, the increased demand for numbers of persons would have been less than the increased demand for man-days, if labor generally were underutilized. However, if the increased demand for man-days were even close to the figure shown in column 6 of Table 24, there would almost surely have been increased demand for workers. If, for example, the average work day was only 75 per cent of a full work day — effective unemployment was 25

---

[31] This range is indicated by the standard error of the regression equation for the 177 farms.

per cent – then a 60 per cent increase in the demand for man-days would result in a 20 per cent increase in the demand for farm workers, all of them fully employed.

The finding that nonfruit farms in O'Higgins could have profitably employed considerably greater quantities of nonland inputs depends crucially upon two features of their operations – the low returns to land and high returns to fertilizers. It is important, therefore, to consider these two features more closely.

Given the marginal product of land in terms of escudos per hectare, the rate of return depends upon the price of land. The reasons for selecting a price of E^O900 per hectare of prime quality land have already been given, and there is not much to be gained by pursuing the matter further. It is worth noting, however, that even if the price were E^O800 per hectare the rate of return would be less than 7 per cent, still well under the assumed opportunity cost of capital.

But is it reasonable to find the rate of return to land so much below the opportunity cost of capital? Most students of Chilean agriculture would probably answer positively, for a variety of reasons. One is that land prices are bid up because land is an effective hedge against inflation. But why should land appear particularly attractive on this score if there are other investment opportunities which also provide a hedge against inflation and earn higher rates of return? It is not obvious, for example, that farm machinery is a less effective hedge against inflation than land. To be sure, inflation will drive up the prices of new machinery so that nominal replacement costs rise rapidly. However, so long as the prices of farm products keep reasonably in step with farm machinery prices, the real replacement cost will not rise appreciably. Of course, the nominal second-hand value of machinery can also be expected to rise. Hence, farm machinery may provide as effective a hedge against inflation as land. In this case, why would land be priced to yield only 6 per cent if machinery investments yield around 15 per cent?

Of course land prices might be expected to rise for reasons that have nothing to do with general inflation. If it were thought that agricultural land would soon be in demand for industrial, commercial, or residential uses, its price could be bid up to a level not justified by its return in agricultural uses. Still the higher price would be "economic" if the expectations were well founded. This phenomenon could perhaps have affected the prices of land in the immediate vicinity of Rancagua and Rengo, but it could not have been very important in the rest of the province. It cannot explain the low returns to land among the sample of farms.

A more plausible explanation is found in the importance of land in obtaining the credits essential for acquiring machinery, fertilizers, and other inputs. The importance of land in establishing credit-worthiness was stressed in the previous chapter, as was the important role of credit in obtaining nonland inputs. In these circumstances it may be completely rational for farmers to make apparently uneconomic investments in land if this strengthens their position in the competition for highly profitable nonland inputs.

This may appear too facile an explanation, however. One may ask why farmers, instead of holding excess land to assure access to credit, did not sell off the excess and use the proceeds to purchase nonland inputs. There is no obvious answer to this question, but a couple of possibilities suggest themselves. Given the prices of land and other inputs in 1958/59, the average farmer would have had to sell off perhaps 10-15 per cent of his land (3-4 hectares of prime land or 6-8 hectares of average quality land) to purchase the more profitable quantities of nonland inputs. It is possible – and in the absence of information this is merely a hypothesis – that the market for such small quantities of land would have been quite inactive. Such purchases would not have appealed to larger farmers unless the plots could have been integrated easily into their existing holdings, a matter of chance, while smaller farmers, because of their limited resources, would have encountered difficulties in financing even these relatively small purchases. A second possibility is that farmers might have considered a transfer of their resources from land to nonland assets as a once-for-all, or at least infrequent, shift in their portfolios. If they held this view and also believed that over the long run nonland investments were riskier than land investments (perhaps because of the unpredictable shifts in trade and price policies affecting nonland inputs in the 1950's), they would have accepted a lower rate of return on land than on nonland assets.

There also were tax advantages in holding land. Agricultural taxes were based on the assessed value of the farm rather than on its income. Because of the failure of assessments to keep up with the inflation, agricultural taxes were very low relative to farm income. It is common knowledge in Chile that persons and firms with nonagricultural interests bought farms and reported nonagricultural income as farm income, thus taking advantage of the low agricultural taxes. How common this practice was is not known, but it is widely believed to have been a significant factor in forcing up the price of farm land.

Thus the low return to land in O'Higgins does not necessarily imply economically irrational behavior on the part of farmers there. It does suggest the possibility of an irrational, or uneconomic, allocation of resources from the standpoint of the country as a whole, however. This would depend upon whether the prices of inputs and outputs faithfully reflected the marginal social values of these resources. If they did so reasonably well, it would have been socially desirable to use some of the resources which went into land to increase the supply of fertilizers and other nonland inputs, to enable nonfruit farmers in O'Higgins to produce more than they did. So long as the availability of these inputs was linked to land ownership, however, this reallocation of resources was not to be expected.

The second major feature of the sample of farms leading to the conclusion that they could have profitably employed considerably more nonland inputs is the high return to fertilizers. Table 24 indicates that an additional escudo spent on fertilizers would have yielded an increment of $E^O4.73$ in output. It is of interest to note that this figure is quite consistent with the results obtained from the analysis in Chapter 3 of the economy as a whole. It was shown that, given plausible assumptions about the relative contribution of fertilizers and pesticides to increased crop production in the 1950's, the marginal product of these inputs was between about $E^O2.00$ and $E^O5.00$.[32] This increases the credibility of the estimated marginal product of fertilizers in O'Higgins and also suggests that the situation there with respect to fertilizers was found generally throughout the country.

The pattern of input use in O'Higgins in the late 1950's is similar in some respects, but not all, to that found in a sample of 81 nonfruit farms in Santiago Province in the early 1960's. In that period the Chilean government, jointly with the Organization of American States, undertook an aerial photography survey of the principal agricultural areas of the country. In connection with this work input-output data were collected from farms in a number of provinces. These farms were not randomly selected, and in some other respects the data collected appeared less satisfactory than those for O'Higgins, with which the present study is principally concerned. Nonetheless, analysis of the more recent data permits some interesting comparisons with the results for O'Higgins.

Table 25 presents the relevant material for 81 nonfruit farms in Santiago Province.[33] They refer to the crop year 1962/1963. Land inputs in this function are measured in hectares adjusted for quality differences in the same way as in the functions for O'Higgins. Labor is measured in costs per man-year, and hence is not directly comparable with the O'Higgins material. Figures for operating expenses, fertilizers and pesticides, and seeds are the expenditures for these inputs in the course of the crop year. Figures for machinery and productive structures are end-of-year values whereas that for livestock is the average of beginning-of-year and end-of-year values.

The regression is highly significant statistically and explains about 86 per cent of the variations in output. Intercorrelations among the independent variables are considerably less than in the O'Higgins data, the maximum correlation coefficient being .79 between livestock and fertilizers. In a few other cases correlation coefficients are between .70 and .75, but for most pairs of inputs they are well under .70.

---

[32] The reader is also reminded that in Chapter 3 the incremental return to fixed capital for the country as a whole was estimated to be between 10 and 15 per cent in the 1950's, figures which are quite consistent with those for nonfruit farms in O'Higgins.

[33] Santiago Province is contiguous on the north to O'Higgins Province. The two provinces are roughly similar in climate and soils.

Table 25. Cobb–Douglas Production Function and Marginal Value Products for 81 Non-fruit Farms in Santiago Province, 1962/63

	Regression coefficients	MVP (escudos)
Land (per hectare, adjusted)	.143[b] (.068)	98.90
Labor	.281[a] (.070)	1.46
Operating expenses except fertilizers, pesticides, and seeds	.123 (.098)	0.80
Fertilizers and pesticides	.157[a] (.046)	4.67
Seeds	.122[a] (.053)	3.21
Machinery	.021 (.077)	0.076
Livestock	−.002 (.022)	−0.011
Productive structures	.086 (.069)	0.289
Sum regression coefficients	.930	
R^2	.859	
A (intercept of production function)	1.082	

Source: Data collected by Ministry of Agriculture in connection with aerial photography survey.

Note: Figures in parentheses are the standard errors of the regression coefficients.

[a]Significantly different from zero at the 1 per cent level of probability in a one-tail test.
[b]Significantly different from zero at the 5 per cent level of probability in a one-tail test.

Since labor input is measured in annual labor costs, the MVP indicates that marginal returns to labor exceeded labor costs by 46 per cent, i.e., and an additional escudo spent for labor added $E^o1.46$ to gross output. By comparison, the MVP of labor in the sample of O'Higgins farms exceeded average labor cost by 14 per cent.

The Santiago MVP's for machinery and productive structures differ rather considerably from those found in O'Higgins. There is no obvious reason for this, and given the large variances in the coefficients for Santiago, there is little point in pushing the comparisons. The MVP for seeds is substantially larger in the Santiago data than in those for O'Higgins. A government program to develop and encourage wider use of improved seed varieties was begun in the late 1950's, and the differences observed between O'Higgins and Santiago may reflect the impact of that program. This cannot be confirmed, however.

The most notable similarities in MVP's for the two areas are those for land, fertilizers, and livestock. Land price data for Santiago are just as scarce as they are for O'Higgins. However, the previously cited index of irrigated land prices[34] increased by about 25 per cent between 1959 and 1962. If we assign a hectare of prime irrigated land in Santiago the same price as in O'Higgins — E⁰900 in 1959 — then the estimated Santiago price in 1962 is E⁰1,125. This is almost surely too low, since land in Santiago Province generally would be priced above that in O'Higgins because of greater proximity to the city of Santiago. Moreover, the relevant land price would be the average for 1962/63, since output refers to that period. Thus the estimate of E⁰1,125 per hectare of prime land is probably quite conservative. Nonetheless, the indicated return at this price is 8.8 per cent, below the minimum assumed opportunity cost of capital.

The MVP's of fertilizers and pesticides were almost the same in Santiago in 1962/63 as in O'Higgins in 1958/59. This may appear somewhat surprising at first glance since fertilizer consumption in the country as a whole increased by 85 per cent from 1958/59 to 1962/63,[35] far more than that of any other major input. In this circumstance, the MVP of fertilizers would be expected to decline, perhaps sharply. Hence one would suppose that the marginal product of fertilizer in both O'Higgins and Santiago would be lower in 1962/63 than in 1958/59. The data presented do not necessarily contradict this, since the marginal product in Santiago in 1958/59 may have been higher than in O'Higgins. But given the proximity of the two provinces and their similarities in soil and climate, it is difficult to see why factor returns should differ widely between them. The anomaly may be more apparent than real, however: the marginal product of fertilizer may have declined little if any despite the sharp increase in consumption. This is suggested by the fact that in the country as a whole, employment of pesticides and farm machinery also rose between 1958/59 and 1962/63, although the increases lagged behind that for fertilizers.[36] The accompanying rise in the use of these inputs would tend to impede the decline in the MVP of fertilizers induced by substantially greater consumption of it.

Thus the returns to land, fertilizers, and livestock in Santiago in 1962/63 were very similar to those in O'Higgins in 1958/59. The returns to labor and

---

[34] Davis, "Precio real de la tierra agrícola."

[35] El uso de fertilizantes. (See Ch. 2, n. 53.)

[36] In 1962/63 pesticide consumption in the country as a whole averaged about 13 per cent above the level of 1958. The stock of farm machinery increased by approximately the same amount. (El uso de pesticidas and El uso de maquinaria. (See Ch. 2, n. 64 and n. 4.)

seeds in Santiago, while different from those in O'Higgins, nevertheless point to a similar situation with respect to patterns of resource use. In general, therefore, the Santiago data indicate that farmers in that province in 1962/63, like those in O'Higgins in 1958/59, apparently could have profitably employed greater quantities of labor, seeds, fertilizers, and productive structures *with no increase in irrigated area.* As in O'Higgins, no increase in the quantity of livestock was indicated.

In his study of productivity on irrigated farms in the Central Valley, Aldunate came to almost identical conclusions with respect to private farms.[37] His data are for the crop year 1963/64. They indicate that the marginal product of land on private farms, regardless of product specialization, was below its opportunity cost, while that for livestock was negative. Marginal returns to fixed capital and operating expenses were above their opportunity costs, while that for labor was slightly below the average wage paid. Aldunate concludes that farmers in the Central Valley could have profitably employed substantially greater quantities of fixed and operating capital with no increase in the quantity of land, while the investment in livestock should have been reduced. He also states that somewhat less labor should have been employed. This conclusion is based on the comparison of the marginal return to labor and its average wage, with all other inputs given at their geometric means. However, if farmers had employed the larger quantities of fixed and operating capital which they could have used profitably, the marginal product of labor would surely have been raised above the average wage. Hence, greater use of capital inputs would have stimulated greater demand for labor.

Finally, a study by Fonck yielded results which in important respects are quite consistent with those presented above.[38] Taking his data from the agricultural census covering operations in the crop year 1954/55, Fonck concentrated his analysis on the region running from Aconcagua Province in the north through the island of Chiloe in the south. This region encompasses Santiago and O'Higgins provinces and accounts for much the greater part of the total agricultural production of the country. Production and input data were available at the commune level. Fonck grouped the data for communes so as to divide the study region into 7 agriculturally homogeneous areas. He then computed production functions for each of the 7 areas as well as for the study region as a whole.

Fonck's inputs are not identical with those employed in this study or in that by Aldunate, so exact comparisons of input marginal products are not possible. Fonck's data apparently included nothing comparable to our input

[37] "A Comparison of Resource Productivity and Efficiency."

[38] Carlos Fonck O., "An Estimate of Agricultural Resource Productivities by Using Aggregate Production Functions," (unpublished M.S. thesis, Cornell University, Ithaca, Feb. 1966).

class labeled "all operating expenses except fertilizers, pesticides, and seeds." The closest thing to a machinery input in Fonck's analysis is an item called "traction force" which is a composite made up of *numbers* of work animals and tractors of various sizes. A weighting system is used to express each component of this input in equivalents of a 2 to 3 plow tractor.

Despite these differences in definitions it is interesting to note that for the area encompassing O'Higgins Province Fonck found:

1. That the marginal returns to land and livestock investments were low and not significantly different from zero;

2. That the marginal return to fertilizer was very high ($E^O6.26$) and clearly different from zero;

3. That the marginal return to labor was high (above the going wage rate) and clearly different from zero.[39]

Fonck's results indicated, therefore, that in the area including O'Higgins Province,[40] output in 1954/55 could have been profitably increased by the employment of more fertilizers and labor, with no increase in the amount of irrigated land. This finding accords well with the conclusions drawn from the analysis of material already presented above.[41]

Thus there is a considerable body of evidence, drawn from a variety of independent sources, in support of the important conclusion that in the 1950's and early 1960's Chilean farmers could have profitably employed substantially greater quantities of most nonland inputs.[42] If this was in fact true, the question naturally occurs, why did they fail to do so? One possible

[39] Fonck's estimates for "traction force" and "non-livestock constructions" were not significantly different from zero at conventional standards of significance. As noted above, he had no input categories for operating expenses other than fertilizers and labor. This summary of Fonck's results is taken from Table 9, p. 36, and Table 19A, p. 62, of his study.

[40] This area accounted for about 22 per cent of total output of the study region.

[41] For the study region as a whole Fonck found the marginal product of irrigated land to be E^O40 per hectare (in prices of 1959). Assuming the price of E^O460 per hectare used above, this indicates a rate of return of about 8.5 per cent, compared with the assumed minimum cost of capital of 10 per cent. The marginal product of fertilizer for the entire study region was $E^O2.03$, while the return to livestock was 32 per cent. The difference between this result for livestock and that for the area including O'Higgins Province may be explained by the fact that the study region as a whole includes major livestock producing areas. O'Higgins is not such an area. Nonetheless, Fonck's high return to livestock is puzzling in view of the generally poor performance of the livestock sector in the country as a whole in the 1950's.

[42] There is a qualification to this statement, however. The percentage of potential expansion for the country as a whole would have been less than the potential for O'Higgins. If farmers throughout the country had attempted to exploit the profit opportunities that evidently existed, the increased demand for inputs would probably have pushed up the prices of at least some of them. At the same time, the very large increase in production would have put downward pressure on product prices. Hence, even if the apparent expansion potential in the country as a whole was similar to that in O'Higgins and Santiago provinces, the real potential would have been less. It would still have been considerable, however.

explanation is that the larger output yielded by employment of more nonland inputs could not have been profitably marketed, either in Chile or abroad. This would imply that the marketing system was operating at or close to the physical limits of capacity at the levels of output achieved in the 1950's. In this case, additional production could not have been marketed at all, or else only with such rapidly rising distribution costs as to eliminate all incentives for increased production.

There is little doubt that marketing of farm products was beset by many problems in the 1950's and subsequently. The CIDA Report asserts that the "marketing structure for agricultural and livestock products is extremely deficient" and that improvement of it must be one of the prime objectives of agricultural policy.[43] A statement by an organization representing the private sector in agriculture also stresses marketing problems,[44] as does the agricultural plan for 1965–80.[45] None of these statements, however, goes so far as to say that because of marketing bottlenecks agricultural production could not have grown faster than it did in the 1950's. Figures in the agricultural plan suggest that marketing problems were not that severe. Investments in storage and handling facilities at all levels of the marketing structure from the farm gate to retail are programed at E^O345 million (1965 prices) in 1968–71. The annual rate, therefore, is E^O86 million. This compares with programed annual investments over this period of E^O120 million in farm machinery and equipment, and total on-the-farm investments of E^O567 million. Expenditures for fertilizers, pesticides, and seeds are projected at E^O247 million annually, and total on-the-farm operating expenses (not including labor) at E^O744 million.[46] Of course, the marketing structure is more than just physical facilities: the organization of activities also is important. Nonetheless, the figures cited suggest that in the second half of the 1960's the expansion of production was restricted more by capacity limitations on the farm than by those in the marketing structure.

Whether this was true also in the 1950's is uncertain. However, there is nothing in the CIDA Report or Plan discussions of marketing bottlenecks to suggest that these limitations were more serious in restricting the expansion of production in the 1950's than in the 1960's. We conclude, therefore, that the failure of farmers to employ more nonland inputs in the 1950's was not due to marketing bottlenecks.

Another possible explanation is that the opportunities for using even greater quantities of these inputs were of recent occurrence and that farmers

[43]CIDA Report, p. 248.

[44]"Los productores y los problemas agropecuarios," statement of the Comisiones de Política Agraria y de Problemas Sociales, Asamblea de Directivas Agrícolas, published in *Panorama económico*, No. 221, Santiago, June 1961, p. 141.

[45]República de Chile, Ministerio de Agricultura, *Plan de desarrollo agropecuario 1965-1980*, Vol. III, Ch. XII (Santiago, 1968).

[46]*Plan de desarrollo agropecuario*, Resumen, pp. 132, 150, and 196.

had not yet had time to react fully to them. For example, if input-output price relations had recently changed favorably, or if improved seeds, fertilizers, or machinery had just become available at favorable prices, farmers would not yet have had time to respond to these new profitable opportunities. The trouble with this explanation is that, with the exception of fertilizers and labor, the prices of agricultural inputs rose persistently faster than product prices in the 1950's.[47] In fact, consideration of relative price changes only deepens the puzzle. If profitable opportunities to use more nonland inputs existed in 1958/59, these opportunities must have been even greater earlier in the decade when relative prices were more favorable to agriculture. Hence, in addition to asking why the quantity of nonland resources was less than the most profitable amount in 1958/59, one must also ask why this underutilization persisted through an entire decade.

It is probably true to say that most students of Chilean agriculture believe that the country's land tenure system holds the answers to both these questions. The argument in support of this explanation has already been presented in Chapter 2 and will not be restated in detail here. The salient points are that the land tenure system deprives the great mass of small farmers of the incentive for technical innovation, and that even if the incentive exists they are denied adequate access to the credit and technical knowledge essential to the economical employment of greater quantities of modern inputs. Moreover, the land tenure system makes the large landowners insensitive to economic stimuli so that even though they have the resources and knowledge needed to innovate, they are not inclined to do so.

In considering this argument in Chapter 2, we concluded that it has considerable validity so far as small farmers are concerned, but that the part dealing with the behavior of large farmers is unconvincing. Nonetheless, the argument still could explain the apparent persistent underutilization of nonland inputs if this were characteristic only of the small farmers. In fact, however, the underutilization of nonland inputs was characteristic of both large and small farms,[48] and, we have argued, the land tenure system cannot explain this in the case of the large farms. Since these farms had 85–90 per cent of the arable land and produced about 60 per cent of total agricultural output in the 1950's, the failure of the land tenure argument to account for their behavior means that it cannot serve as a general explanation of the performance of Chilean agriculture in that decade.[49]

[47] See Table 2 above.

[48] The 177 sample farms were classified by size groups and production functions calculated for those in the 1–50 hectare category as well as for those with 150–200 hectares. The results showed that returns to land were relatively low on both groups of farms but returns to fertilizers and pesticides were very high. The larger farms also could have profitably employed considerably more labor and machinery, whereas the smaller farms could have used more seeds and productive structures.

[49] The percentages on amount of land and production of large farms are from the CIDA Report, p. 206.

There is another hypothesis, however, which is consistent with underutilization of nonland resources on both large and small farms, namely, that the use of these resources was subject to various forms of nonprice rationing imposed by foreign-trade and credit policies. If this hypothesis is correct we would expect to find (1) that imports of farm inputs were subject to various governmental controls and were responsive to the availability of credit; (2) that import and credit policies in fact restricted the ability of farmers to acquire these inputs in the amounts they wished; (3) that the prices of these inputs did not rise so rapidly in response to the strong demand for them in the 1950's as to eliminate their profitability.

There is ample evidence that these conditions were satisfied throughout the decade of the 1950's. It was shown in Chapter 2 that almost all farm machinery and a substantial proportion of pesticides and phosphate fertilizers – the type which accounted for the bulk of the increase in fertilizer consumption in the decade – were imported. It also was shown that because of the country's chronic balance of payments problem in this period these imports were subject to a variety of direct and indirect controls – a formal request for permission to import had to be made and acted on, substantial deposits were required in advance of importation, and so on. Moreover, it was shown that a major proportion of purchases of farm machinery, fertilizers, and pesticides were financed with credit. In the case of machinery it was possible to demonstrate this relationship statistically. In fact, credit policy alone may be sufficient to explain the behavior of machinery imports in the 1950's. Between 1950 and 1955 credits extended by CORFO and the State Bank (the only significant sources of machinery credits) tripled and imports of farm machinery increased from $2.5 million to $17.0 million. From 1955 to 1959, credits declined by 55 per cent, largely because CORFO made no machinery loans in this period, and machinery imports fell to $5.5 million.

The effects of credit, and trade, policies in restricting purchases of fertilizers and pesticides cannot be so directly observed as in the case of farm machinery. As noted in Chapter 2, fertilizer credits extended by the State Bank were about 65 per cent of total fertilizer purchases in the early 1960's. While data on the Bank's fertilizer credits in the 1950's are not available, the real value of the Bank's total agricultural loans declined over the course of the decade.

This was true of total agricultural credits, as was demonstrated in Chapter 2. In principle, of course, the decline in the real value of credits could have resulted from weak demand. In fact, however, there is little doubt that the decline was caused by restriction on credit supply. It was noted in Chapter 2 that particularly after the mid-1950's the Central Bank adopted policies designed to limit the expansion of credit, and that the real value of credits extended to the whole private sector – not just agriculture – declined over the decade. Other evidence also suggests a shortage of agricultural credit. In a report dealing with the years 1951–62 it was concluded that "the total demand for agricultural credits greatly exceeds the financial capacity of the lending institutions."[50] The report refers to a study by the Agricultural Department of the State Bank which showed that over a four-year period in the 1950's (years not given) 25–40 per cent of the loan applications to the Department were turned down.[51]

The hypothesis under consideration can also be evaluated by observing the behavior of imports of machinery, fertilizers, pesticides, and other farm inputs. If there were a large potential demand for these inputs which could not be realized because of restrictive credit and import policies, then we would expect imports of the inputs to rise significantly when the policies became less restrictive. This in fact happened. After 1959 both credit and import policies were eased, and imports of farm machinery, fertilizers, pesticides, and other inputs rose sharply. (See Table 26.) The real value of bank loans to agriculture in 1961–65 was 69 per cent above the level of 1956–60.[52] Import controls also were loosened. In May 1960 farm machinery and spare parts were put in the category requiring the lowest deposit percentage (5 per cent), and in subsequent years the system of prior deposits was done away with altogether. This is not to say that all import controls were abandoned; however, there was a significant liberalization of import policies applicable to farm inputs.

The rationing of farm inputs by foreign-trade and credit policies would not in itself explain the persistence over the decade of profitable opportunities to use more of these inputs. If their prices had risen freely in response to the excess of demand over supply, the margins of profitability would soon have been eliminated. Hence the credibility of the hypothesis requires evidence that prices of these inputs were not fully responsive to the excess of demand over supply. For machinery, phosphate fertilizers, and to a lesser extent pesticides, prices to farmers have two components: (1) the c.i.f. import price set basically by world market forces and Chilean foreign-exchange policy; (2) marketing margins in Chile, including profits of importers and distributors. Since the demand in Chile for these inputs is a relatively small part of total world demand, their c.i.f. prices expressed in foreign currency would not be significantly influenced by market conditions in Chile. The conversion of these prices to escudos of course depends upon the rate of exchange. The explanation of Chilean foreign-exchange policy is not within the bounds of this study. However, it clearly was not managed so as to permit the escudo price of farm inputs to rise in response to the demand for them. Indeed, as was pointed out in Chapter 2, the multiple-exchange-rate system in effect until the mid-1950's actually held the price of farm machinery below the level it would have reached with a freely fluctuating exchange.

[50]Consejo Superior de Fomento Agropecuario (CONSFA), *El crédito agrícola durante el período 1951–1962* (Santiago), p. 27.

[51]Ibid.

[52]ODEPA, unpublished memorandum.

Table 26. Imports of Agricultural Machinery, Fertilizers, Pesticides, and Other Inputs

Year	Machinery (million current U.S. $)	Fertilizers (million U.S. $, 1965 prices)	Pesticides (million U.S. $, 1965 prices)	Other nonmachinery inputs (million U.S. $, 1965 prices)
1950	2.5	–	–	–
1951	8.0	–	–	–
1952	7.2	–	–	–
1953	9.8	–	–	–
1954	14.0	–	–	–
1955	17.0	2.9	0.1	2.7
1956	9.4	2.1	0.6	3.5
1957	11.9	1.3	0.6	2.2
1958	7.1	3.6	0.5	1.8
1959	5.5	4.0	1.2	1.2
1960	11.2	5.2	0.8	1.7
1961	15.6	8.1	1.0	1.7
1962	13.2	13.3	0.6	1.2
1963	12.7	17.7	2.0	2.1
1964	9.5	15.6	1.7	2.3
1965	11.7	15.7	1.9	2.1

Sources: Machinery—Ch. 2, Table 5; all others—ODEPA unpublished memo.

Note: Data for fertilizers, pesticides, and other nonmachinery inputs not available until 1955.

Hence, it seems permissible to conclude that the c.i.f. escudo prices of farm machinery, phosphate fertilizers, and pesticides were not responsive to the excess of demand over supply of them. Internal marketing margins, however, might have been expected to rise in response to the excess demand if importers and distributors were able to raise prices freely. We know from Chapter 2 that prices of machinery, fertilizers, pesticides, and other nonlabor inputs rose in the 1950's; in fact they rose more rapidly than prices of farm products. Nonetheless, the prices of most of these inputs were subject to controls in this period, which probably prevented them from rising as rapidly as they would have in the absence of controls. Thus there is general evidence that importers and distributors of these inputs were not able to increase prices freely in response to market conditions.

A more detailed examination of the marketing margins of these inputs in the 1950's is not within the scope of this study. In any case, it appears unnecessary for present purposes. As noted earlier, CORFO and the State Bank were major factors in the importation and distribution of farm machinery and fertilizers. The marketing margins they charged were in general designed to cover actual costs of distribution, taxes, and a fixed profit percentage.[53] Hence these agencies did not try to capture the profits implicit in the excess of demand over supply of these inputs. Indeed, in the case of

[53]This statement is based upon the analysis of marketing margins for fertilizers and farm machinery given in El uso de fertilizantes and El uso de maquinaria, respectively.

fertilizers a subsidy was applied which throughout much of the 1950's held the price of fertilizers to farmers below the levels it otherwise would have reached.

There is considerable evidence, therefore, for the belief that the persistence in the 1950's of unexploited opportunities for the greater use of fertilizers, machinery, and pesticides was due to government foreign-trade and credit policies, which imposed a form of nonprice rationing on the supply of these inputs. Except for these policies, employment of the inputs probably would have been substantially greater and agricultural production correspondingly larger. Thus these policies cost the country something – perhaps a great deal – in agricultural production foregone.[54] This does not necessarily mean that these policies were wrong. Such a judgment would require the assumption that the prices of the various inputs correctly measured their social costs, and we are not prepared to make that assumption. For example, the foreign exchange which would have been required to import larger quantities of machinery and fertilizers may have had a larger social pay-off in the uses to which it was actually put. Or the policy of credit restraint may have been necessary to prevent Chile's endemic inflation from literally "running away," with consequent social and economic losses which would have been substantially greater than the loss in agricultural production caused by pursuit of this policy.

These examples are cited only to show that the foreign-trade and credit policies which prevented agricultural production from realizing its full potential may have been justifiable in terms of overall social, economic, and political objectives of the country. Whether this was in fact the case cannot be determined in this study. Any such determination, however, would have to include an evaluation of the cost of these policies in terms of foregone agricultural production. The evidence presented here suggests that this cost may have been high.

Concluding Comments

The emphasis placed on the role of input supply limitations in restricting the growth of output in the 1950's in no way denies that Chilean agriculture also suffered other serious limitations. In the discussion above of the land tenure system it was noted that the system discriminated strongly against small farmers and farm workers in the competition for "modern" inputs and, in the case of renters or leaseholders, weakened their incentives to innovate. Moreover, the agricultural marketing system contained bottlenecks owing to inadequate storage, processing, and transport facilities; price policies were

---

[54]It was noted in Ch. 3 that the rate of growth in crop production in the second half of the 1950's was considerably less than in the first half, and that this matched the pattern of expansion in the stock of farm machinery and in the consumption of fertilizers and pesticides.

subject to frequent and unpredictable changes; and tax policies may have contributed to inefficient use of agricultural resources.[55] If some or all of these conditions had been improved, *and if the supply of imported farm inputs had been permitted to increase freely*, then no doubt production would have increased much more rapidly than it did in the 1950's. If these conditions had been improved *without* a more ample supply of imported inputs, however, the impact on production in the decade probably would have been small.

Finally, let it be noted that in the author's judgment the conclusion just stated in no way contradicts the idea that to achieve healthy development of its agriculture over the long run, Chile needs to reform its land tenure system, expand and modernize the marketing and transport infrastructure serving agriculture, rationalize agricultural pricing policies, and adopt an agricultural tax system which rewards efficiency.

[55] As noted earlier in this chapter, the tax system encouraged persons and firms with nonagricultural interests to acquire farms. Of course it would not have been to the advantage of these persons or firms to deliberately manage their farms inefficiently; however, since their main interests lay outside agriculture, management efficiency may have suffered.

Patterns of Resource Use
and Productivity

The preceding chapter dealt with patterns of resource use in Chilean agriculture in the 1950's and concluded that apparently large untapped opportunities to expand production persisted throughout the decade. Untapped potential production was defined (implicitly) as the difference between actual production and the amount that could have been profitably produced, had the necessary inputs been available in sufficient quantity.

There is another way of defining potential production which also offers promising insights into agricultural performance. Potential in this case is represented by the differences in total productivity that exist between farms. It is a common observation that two farms may have substantial and persistent differences in output even though both employ virtually identical quantities of what appear to be the same kinds of inputs. The difference between outputs obtained from a given quantity of inputs defines the difference in productivity between the two farms. *Measurement* of this difference gives an indication of the gain in output to be had by raising the productivity of the less productive farm to equal that of the more productive farm. *Explanation* of the difference will suggest the measures which must be taken to achieve the higher level of productivity.

CONCEPT AND MEASUREMENT OF PRODUCTIVITY

In the discussion so far, the word "productivity" has been used as though it had one generally understood and accepted meaning. This is not the case, however. Of course, everyone knows that productivity refers to the amount of output per unit of input; but there are many possible ways of defining inputs, and the measurement of productivity will depend upon the definition selected.[1] Frequently one encounters measures of productivity stated in terms of output per unit of some single input, such as capital, or labor, or land. The last is frequently encountered in the case of agriculture. It is common to find statements that one area is "more productive" than another area because output per hectare is higher in one than in the other.

Such assertions may be quite misleading, however, if they are unaccompanied, as is usually the case, by statements about the quantities of other inputs combined with land to produce the observed output. Assume a situation in which two areas each use only two inputs, land and labor. Area A produces 25 per cent more per hectare than Area B, but uses five times as much labor per unit of output. Area A is thus 25 per cent more productive per unit of land, but Area B is 400 per cent more productive per unit of labor. Taking into account all inputs in each area, which area is more productive?

It will be shown below that there is usually no unambiguous answer to this question. However, the area of ambiguity can be reduced by construction of a productivity measure which takes account of the separate contribution of each input to output. There are a variety of ways of doing this. The one selected in this study is to calculate the marginal value products of different inputs on groups of farms and combine these in an index which measures the productivity of one group in relation to the other group. In the case of the farms of Areas A and B suppose that,

MP_{L_a} = marginal value product of labor in Area A

MP_{T_a} = marginal value product of land in Area A

MP_{L_b} = marginal value product of labor in Area B

MP_{T_b} = marginal value product of land in Area B

L_a = quantity of labor in Area A

T_a = quantity of land in Area A

L_b = quantity of labor in Area B

T_b = quantity of land in Area B

P_a = productivity of Area A in relation to Area B

---

[1] There are also alternative measures of output, depending upon the extent to which it is net of certain inputs. Net national product, for example, is "net" of gross national product by the amount of capital consumption. Output at the farm level may be defined as gross farm output, i.e., total sales plus home consumption plus inventory change, or it may be defined net of certain purchases such as fertilizers or seeds. The difficult problems in the measurement of productivity, however, arise on the input side.

The combined productivity of land and labor in Area A can be compared with that in Area B by means of the ratio

$$P_a = \frac{L_a MP_{L_a} + T_a MP_{T_a}}{L_a MP_{L_b} + T_a MP_{T_b}} \tag{1}$$

The numerator in this ratio is actual output in Area A; that is, total output in A is equal to the sum of the products of input quantities times input marginal products.[2] The denominator indicates the amount that would have been produced in Area A, given its complement of inputs, if the marginal product of each input had been the same as in Area B. Hence if, in combination, the marginal products of A's inputs were higher than those of B's inputs, the ratio would have a value greater than 1, the amount of the excess indicating the per cent by which A's productivity exceeds that of B's.

By taking into account the contribution to production of both inputs, this procedure clearly provides a better comparison of productivity in Areas A and B than any single input measure can do. The procedure can be extended to encompass as many inputs as are included in the production process, or for which adequate measures of quantity can be found.

Even if all inputs could be properly quantified and their marginal contributions to output correctly measured, the above procedure will usually yield ambiguous results. Ambiguous here means that the procedure will not provide a unique measure of relative productivity differences between groups of farms. The reason can be made clear by returning to the example of Areas A and B. Suppose that in A labor is cheaper in relation to land than it is in B. Farmers in A, therefore, will tend to use more labor per unit of land than farmers in B. The quantities of land and labor in the two areas might appear as follows:

$$L_a = 150$$
$$T_a = 100$$
$$L_b = 100$$
$$T_b = 100$$

Then the productivity ratio could be written

$$P_a = \frac{150\, MP_{L_a} + 100\, MP_{T_a}}{150\, MP_{L_b} + 100\, MP_{T_b}} \tag{2}$$

However, we can also compare the two areas by calculating the ratio of Area B's actual production to what it would have produced had the marginal

---

[2] The procedure thus assumes constant returns to scale.

value products of its inputs been the same as those for Area A. This ratio (P_b) is written

$$P_b = \frac{100\,MP_{L_b} + 100\,MP_{T_b}}{100\,MP_{L_a} + 100\,MP_{T_a}} \tag{3}$$

For the procedure to give a unique measure of relative productivity differences between the two areas it is necessary that the reciprocal of equation (3) equal equation (2). That is, it must be true that

$$\frac{100\,MP_{L_a} + 100\,MP_{T_a}}{100\,MP_{L_b} + 100\,MP_{T_b}} = \frac{150\,MP_{L_a} + 100\,MP_{T_a}}{150\,MP_{L_b} + 100\,MP_{T_b}} \tag{4}$$

The statement obviously is not true.[3] Hence there is no unique measure of the productivity difference between the two areas. Instead, the amount of the difference will depend upon which area's inputs are used as weights in constructing the productivity index.

Note that for (4) to be true it is not necessary that the absolute amounts of inputs be the same in A and B. Suppose that, instead of the previously assumed amounts, the quantities of land and labor in Area B are 200 and 300 respectively. In this case (4) would be written

$$\frac{300\,MP_{L_a} + 200\,MP_{T_a}}{300\,MP_{L_b} + 200\,MP_{T_b}} = \frac{150\,MP_{L_a} + 100\,MP_{T_a}}{150\,MP_{L_b} + 100\,MP_{T_b}} \tag{4a}$$

which is a true statement. The essential condition for securing a unique measure of productivity differences with this procedure is that the *proportions* between inputs in the two areas be the same.[4] The problem is that this

[3] Except in the uninteresting case in which the marginal products of corresponding inputs are identical in both areas.

[4] The reader will no doubt have observed by now that we deal here with a form of the index number problem. The productivity ratios are indexes in which the quantities of inputs employed on one or the other group of farms are used as weights. As in any index number, variations in weights have no effect on the value of the index so long as the proportions between the weights are constant. If these proportions change, however, different index values will be obtained. The problem also affects measurements of productivity over time. If, because of technical progress, or any other reason, the productivity of some inputs increases faster over time than the productivity of others, more of the former will be used in relation to the latter. A productivity index based on beginning-period weights will therefore have a different value than an index based on end-of-period weights.

will seldom be the case, and for a very good reason. Farmers will tend to combine inputs in accordance with their relative scarcities, using less of those in relatively short supply and more of those whose supply is ample. If resources were perfectly mobile and all the other conditions of orthodox price theory were met, resources would move in response to differences in relative scarcities, so that there would be a tendency for the prices for given kinds of inputs to be equalized. Farmers everywhere would therefore combine resources in the same proportions; however, the measurement of productivity differences would then be irrelevant, since there would not be any. For if all the above conditions were met, given kinds of inputs would be used in proportions such as to equalize their marginal products, so that productivity differences between farms would be eliminated. The existence of such differences is testimony to the fact that we live in a world other than that postulated in orthodox price theory. That is to say, resources in fact are not perfectly mobile and the other conditions of supply of inputs may also vary from farmer to farmer. It must be expected, therefore, that even within a reasonably small geographical area different farmers will choose different input combinations.

The severity of this problem for the measurement of productivity differences depends upon the extent to which input proportions differ. In the present study they are not so great as to produce widely different measures of productivity. Consequently, the procedure described above has been adopted for use here. Henceforth when reference is made to productivity, or productivity differences, these terms will refer to ratios like that in (1). The relation between output and any one input will be referred to as output per unit of that input.

The productivity index employed here bears a strong similarity to that developed by Kendrick.[5] The principal differences are two: (1) Kendrick is concerned with changes in productivity over time, whereas we are interested in differences between producers in a given period of time; (2) Kendrick uses the prices of inputs as a measure of their marginal productivity whereas we use the marginal value products themselves.[6] We chose to concentrate on

---

[5] John W. Kendrick, *Productivity Trends in the United States*, National Bureau of Economic Research (Princeton: Princeton University Press, 1961).

[6] Kendrick's index can be illustrated for the simple case of two inputs, capital and labor. Let

K_1 = quantity of capital in the given period

K_o = quantity of capital in the base period

L_1 = quantity of labor in the given period

L_o = quantity of labor in the base period

P_{K_1} = "price" (i.e., marginal per cent rate of return) of capital in the given period

P_{K_o} = "price" of capital in the base period

P_{L_1} = "price" (i.e., the wage rate) of labor in the given period

interfarm differences in productivity rather than on changes over time because the data required to concentrate on the latter were not available. We employed marginal value products rather than input prices because the assumption that the two were the same was simply untenable in the face of strong evidence that factor markets work quite imperfectly in Chile, and that the conditions of access to these markets differ considerably between farmers. A secondary but not negligible consideration was that price information for farm inputs is scarce and frequently unreliable.

The productivity index employed here compares the actual output of a group of farms with its hypothetical output, i.e., the amount the group would have produced had the marginal products of its inputs been the same as those of the group with which the comparison is made. It may be objected that neither group could have attained the marginal products of the other group, given each group's complement of inputs, because the marginal products in each case are in part a function of the quantities of inputs employed, and the procedure takes these quantities as given. Indeed, if farmers behave rationally, they will choose quantities of each input in such a way as to maximize their income, given the constraints under which they operate. Hence, given these constraints for each farm group, neither could have, nor indeed should have, employed inputs in quantities any different from those actually chosen.

P_{L_o} = "price" of labor in the base period

Q_1 = total output in the given period

Q_o = total output in the base period

A_1 = total productivity in the given period in relation to the base period.

Kendrick assumes constant returns to scale. Hence the amount of labor multiplied by its price plus the amount of capital multiplied by its price is equal to total output. That is, for any period,

$$LP_L + KP_K = Q.$$

Assuming that factor prices are equal to their marginal products, Kendrick's index can be written

$$A_1 = \frac{\dfrac{Q_1}{Q_o}}{\dfrac{L_1}{L_o}\left(\dfrac{L_o P_{L_o}}{Q_o}\right) + \dfrac{K_1}{K_o}\left(\dfrac{K_o P_{K_o}}{Q_o}\right)} = \frac{\dfrac{Q_1}{Q_o}}{\dfrac{L_1 P_{L_o} + K_1 P_{K_o}}{Q_o}} = \frac{L_1 P_{L_1} + K_1 P_{K_1}}{L_1 P_{L_o} + K_1 P_{K_o}}.$$

Kendrick's index, therefore, compares actual output in the given period with what it would have been, given the quantities of each input in that period, if the productivity of the inputs had been the same as in the base period. In the example, given year input quantities are used as weights. However, the index could also be computed using base year weights. For reasons noted above, the two indexes would differ if inputs were combined in different proportions in the given and base periods.

These statements are correct, but they reflect a misunderstanding of the purpose of the index. That purpose is solely to measure relative productivity differences. In itself the index has no normative significance whatsoever; it is in no sense a measure of the difference between what farmers actually produced and what they should have produced in order to have been good farmers. The question of why productivity differences existed is quite apart from the measurement of these differences. To be sure, farmer quality differences may be a big part of the explanation. But they are not assumed to be the whole explanation. Calculation of the productivity index, therefore, in a sense merely sets the stage for the important part of the study – analysis of the reasons why groups of farms differ with respect to productivity.[7]

Another possible objection to the proposed productivity index must be considered. Suppose that all farms are on the same production function but that because of inefficient management, indolence, lack of credit, or whatever, some farmers use less than the optimum amounts of inputs. Their marginal value products will then be above the prices of these inputs. (This is the meaning of less than optimum amounts of inputs.) Efficiently managed farms with adequate credit, on the other hand, will employ inputs to the point where their marginal value products and prices are equal. Hence, under the assumptions, efficient, well-managed farms will have lower marginal value products and appear to be less productive than inefficient, poorly managed farms! In this case the productivity ratio surely is misleading.

This situation perhaps is theoretically possible, but it seems highly unlikely. In the first place, farms on a given production function cannot use less than the optimum amounts of *all* inputs. Given the qualities of inputs, their marginal value products depend upon the relative amounts employed. To assert that less than the optimum amount of one input is used (its MVP is high) means that not enough of it is employed in relation to the quantities of other inputs. As more of the underutilized input is used its MVP will decline while those of the other inputs will rise. Hence, to assert that less than the optimum amount of one input is employed implies that one or more other inputs are used to excess in relation to the first input. Thus while on ineffi-

----

[7]A limitation of the index employed here is that, like Kendrick's, it implies constant returns to scale. This turned out to be a minor problem because, with one exception reported on below, all the farm groups studied operated with production functions which closely approximated constant returns. That is, in the Cobb-Douglas production functions for each farm group the sums of the regression coefficients were not significantly different from 1. Thus there was no serious violation in assuming that the various farm groups in fact operated with constant return to scale.

A productivity index could also be constructed by substituting the inputs for one group of farms in the production function for another group and dividing the result into the first group's actual output. This procedure is very similar to that adapted here in that it compares the actual output of a group of farms with what the group would have produced, given its complement of inputs, if it had utilized the same production function as the second group of farms. For the case of constant returns to scale, this procedure is formally identical to that employed here.

cient farms the MVP of the first input will be higher than that of the corresponding input on efficient farms, the MVP's of other inputs will be lower on the inefficient farms.

In the hypothetical situation postulated, therefore, the proposed productivity ratio would not necessarily show inefficient farms to be "more productive" than efficient farms, although it might. The outcome would depend upon whether the high MVP's of the relatively underutilized resources more than offset the low MVP's of the relatively overutilized resources. A priori this difference could go either way. In the actual cases considered here, however, it is safe to say that this problem does not arise. In the first place, the more productive groups of farms also are generally more efficient, i.e., the differences between the MVP's and opportunity costs of their inputs are generally less than these differences on the less productive farms.[8] Secondly, with a few minor exceptions, the MVP's of *all* inputs on the more productive farms are higher than those on the less productive ones. As just noted, this would be impossible if the high MVP's simply represented inefficient use of resources. For these reasons it is believed that the hypothetical problem dealt with here is not a real one in the specific cases treated below. This does not mean that some of the differences between marginal value products on more and less productive farms may not reflect differences in allocative efficiency. It does mean, however, that the technique employed for measuring productivity differences is most unlikely to yield perverse results in the sense of showing really less productive farms to be more productive and vice versa.

CALCULATION OF PRODUCTIVITY DIFFERENCES

Four sets of productivity indexes were calculated, one for the entire sample of 177 nonfruit farms and one for each of the three groups obtained when these farms are classified by product specialization (or lack of it). The steps in the procedure were as follows:

1. The inputs of *each farm* in each group were multiplied by the corresponding marginal products derived from the function for its group. In the case of all nonfruit farms the marginal products were those for the 177 farms depicted in Table 21. For the three specialization groups the marginal products were those shown in Table 22.[9] The products obtained from multiplying marginal products by inputs were added. This sum represents the "hypothetical" output of each farm. It is the amount the farm would have produced,

---

[8] See Tables 27 through 30. Opportunity costs are those assumed in Table 23.

[9] As noted above, the procedure assumes constant returns to scale. Consequently, in those functions for which the sum of the regression coefficients exceeded or fell short of 1, the marginal products derived from the function were divided by the sum of the regression coefficients before they were employed in the productivity index.

given its complement of inputs, had the marginal products of its inputs been the same as those for all farms in its specialization group.

2. Actual output of each farm was divided by its hypothetical output. If the result was greater than 1, the farm was more productive than average, while a result less than 1 indicates less than average productivity.

3. From among the 177 nonfruit farms the 60 most productive and 60 least productive were selected in accordance with the productivity ratios calculated in step 2. The farms in the three specialization groups were similarly divided into most productive and least productive groups.

4. Production functions were calculated for each group of most productive farms (MPF) and least productive farms (LPF) and marginal products derived for each input. These functions and marginal products are shown in Tables 27 through 30.

5. As before, marginal products were scaled up or down in proportion to the difference between the sum of regression coefficients and 1. Then for the nonfruit farms and for each specialization category the average (geometric mean) amount of each input on LPF was multiplied by the marginal product of the corresponding input on MPF and the results were added. This sum (hypothetical production) is the amount LPF would have produced had their inputs had the same marginal products as those of MPF.

6. Actual (geometric mean) production of LPF was divided by their hypothetical production. The resulting ratio expresses the productivity of LPF relative to that of MPF.

7. Steps 5 and 6 were repeated, except that MPF inputs were multiplied by LPF marginal products to obtain hypothetical output of MPF.

Before considering the productivity ratios yielded by this procedure it may be well to deal with a question which by now may be disturbing the reader. When one calculates a production function the usual assumption is that the scatter of actual observations around the regression "line" describing the average relation between inputs and outputs reflects merely random disturbances. The procedure adopted here, on the contrary, assumes that the differences between farms lying above and below the regression "line" are due to genuine (i.e., persistent, nonrandom) differences in total productivity. Another way of putting it is that the procedure employed assumes that in fact the above average productivity farms lie on a "higher" production function than the farms of below average productivity. Can this procedure be reconciled with the usual assumption underlying regression analysis?

Ultimately the answer must depend upon the results obtained from this procedure. If the magnitude of the productivity differences between the MPF and LPF appear plausible and, of particular importance, if the two groups differ with respect to characteristics which on both theoretical and empirical grounds are believed to be associated with differences in productivity, then

Table 27. Production Functions and Other Information for 60 Most Productive and 60 Least Productive Nonfruit Farms in O'Higgins Province, 1958/59

Inputs	Regression coefficients MPF	Regression coefficients LPF	Marginal value products (E°) MPF	Marginal value products (E°) LPF
Land, adjusted (hectares)	.208[a] (.061)	.305[a] (.059)	72	43
Labor (man-days)	.324[a] (.063)	.205[a] (.068)	1.72	0.45
All operating expenses except fertilizers, pesticides, and seeds (E°)	.242[a] (.039)	.319[a] (.045)	1.67	1.20
Seeds (E°)	.071[c] (.046)	.054[b] (.027)	1.66	0.76
Fertilizers and pesticides (E°)	.046[a] (.017)	.071[a] (.018)	5.05	4.50
Machinery (E°)	.062[a] (.024)	.032 (.030)	0.22	0.06
Livestock (E°)	.018 (.015)	.031 (.037)	0.08	0.06
Productive structures (E°)	.014 (.013)	.016 (.013)	0.06	0.08
A (intercept of production function)	.172	.078		
Sum regression coefficients	.984	1.032		
R^2	.99	.98		
Average output (E°): Geometric mean	12,500	2,390		
Arithmetic mean	26,328	7,243		
Average size (hectares of prime quality land): Geometric mean	36.6	16.5		
Arithmetic mean	77.0	38.0		

Notes:
Numbers in parentheses are standard errors of the regression coefficients.
On MPF, the regression coefficients for livestock and productive structures were significantly different from zero at 13 per cent and 14 per cent probability levels, respectively, in a one-tail test.
On LPF, this same test showed that the regression coefficients for machinery, livestock, and productive structures were significantly different from zero at the 15 per cent, 20 per cent, and 11 per cent probability levels, respectively.

[a]Regression coefficients significantly different from zero at the 1 per cent level of probability in a one-tail test.
[b]Regression coefficients significantly different from zero at the 2.5 per cent level of probability in a one-tail test.
[c]Regression coefficients significantly different from zero at the 6 per cent level of probability in a one-tail test.

Table 28.   Production Functions and Other Information for Most Productive and Least Productive Livestock Farms in O'Higgins Province, 1958/59

	Regression coefficients		Marginal value products (E°)	
	MPF (20 farms)	LPF (21 farms)	MPF	LPF
Land, unadjusted (hectares)[a]	.069 (.161)	.120[c] (.053)	12.05	13.46
Labor (man-days)	.067 (.176)	.035 (.081)	0.35	0.12
Total operating expenses (E°)	.402[b] (.091)	.546[b] (.041)	1.66	1.24
Livestock (E°)	.256[c] (.122)	.140[b] (.027)	0.50	0.28
Machinery and productive structures (E°)	.166[c] (.079)	.189[b] (.034)	0.23	0.19
A (intercept of production function)	.350	.144		
Sum regression coefficients	.960	1.030		
R^2	.990	.998		
Geometric mean output (E°)	6,970	3,081		
Average size (hectares): Geometric mean	39.9	27.5		
Arithmetic mean	114.2	66.1		

Note: Numbers in parentheses are standard errors of the regression coefficients.

[a]Use of adjusted land was not feasible in the function for this or either of the other two specialization groups because there were not enough farms for which this adjustment could be made.

[b]Regression coefficients significantly different from zero at 1 per cent probability in a one-tail test.

[c]Regression coefficients significantly different from zero at 5 per cent probability in a one-tail test.

the procedure would appear legitimate, despite the violation of the assumption underlying regression analysis. The reader therefore is asked to defer judgment on this point until the analysis has been developed and the results have been presented, which is the main task of this chapter. Let us now turn to it.

ANALYSIS OF PRODUCTIVITY DIFFERENCES

The extent of the productivity differences between the various groups of MPF and LPF is indicated in Table 31. The productivity ratios in the table

Table 29. Production Functions and Other Information for Most Productive and Least Productive Annual Crop Farms in O'Higgins Province, 1958/59

	Regression coefficients		Marginal value products (E°)	
	MPF (37 farms)	LPF (37 farms)	MPF	LPF
Land, unadjusted (hectares)	.195[b] (.092)	.303[a] (.086)	37.77	29.12
Labor (man-days)	.381[a] (.090)	.344[a] (.099)	1.98	0.98
All operating expenses except fertilizers, pesticides, and seeds (E°)	.263[a] (.062)	.373[a] (.058)	2.00	1.66
Seeds (E°)	.081[c] (.057)	.046[c] (.030)	1.73	0.63
Fertilizers and pesticides (E°)	.052[b] (.026)	.018 (.026)	5.02	0.83
Machinery (E°)	.041 (.033)	.030 (.043)	0.16	0.06
Livestock and productive structures (E°)	−.001 (.045)	−.017 (.064)	−0.00	−0.02
A (intercept of production function)	.0106	−.3660		
Sum regression coefficients	1.012	1.097		
R^2	.980	.974		
Geometric mean output (E°)	11,110	4,509		
Average size (hectares):				
Geometric mean	57.5	47.0		
Arithmetic mean	97.8	75.0		

Note: Numbers in parentheses are standard errors of the regression coefficients.

[a]Regression coefficients significantly different from zero at 1 per cent level of probability in a one-tail test.
[b]Regression coefficients significantly different from zero at 5 per cent level of probability in a one-tail test.
[c]Regression coefficients significantly different from zero at 10 per cent level of probability in a one-tail test.

were obtained through the procedure described above. The meaning of these ratios can be made clear through an example. The 60 most productive non-fruit farms actually produced an average (geometric mean) of $E^\circ 12,500$. Had the marginal products of MPF been the same as those of LPF, then with the quantities of inputs they actually used, MPF would have produced $E^\circ 6,393$. The difference between this figure and actual production reflects the differences between the marginal products of MPF and LPF. On balance this differ-

Table 30. Production Functions and Other Information for Most Productive and Least Productive Unspecialized Farms, O'Higgins Province, 1958/59

	Regression coefficients		Marginal value products (E^o)	
	MPF (32 farms)	LPF (32 farms)	MPF	LPF
Land, unadjusted (hectares)	-.078 (.096)	.098 (.128)	-10.97	7.85
Labor (man-days)	.467[a] (.102)	.420[a] (.107)	1.91	1.06
Fertilizers and pesticides (E^o)	.073[b] (.034)	.192[a] (.031)	11.19	11.42
Machinery (E^o)	.012 (.046)	-.101[b] (.050)	0.04	-0.23
Livestock (E^o)	.140[b] (.060)	.091[c] (.063)	0.36	0.18
Productive structures (E^o)	.014 (.014)	.013 (.019)	0.12	0.07
All operating expenses except fertilizers and pesticides (E^o)	.386[a] (.074)	.339[a] (.103)	1.50	0.94
A (intercept of production function)	-.224	-.274		
Sum regression coefficients	1.014	1.052		
R^2	.987	.987		
Geometric mean output (E^o)	5,610	3,417		
Average size (hectares):				
Geometric mean	39.9	42.8		
Arithmetic mean	80.6	32.1		

Note: Numbers in parentheses are standard errors of the regression coefficients.

[a]Regression coefficients significantly different from zero at 1 per cent level of probability in a one-tail test.
[b]Regression coefficients significantly different from zero at 5 per cent level of probability in a one-tail test.
[c]Regression coefficients significantly different from zero at 10 per cent level of probability in a one-tail test.

ence was 96 per cent, i.e., the marginal products of MPF were, on average, 96 per cent higher than those of LPF.

As noted above, these ratios are indexes of productivity in which the weights are the quantities of inputs employed on one or the other of the two groups of farms. In the example just cited the weights are the inputs of the MPF. If the inputs of the LPF are used as weights a different productivity ratio will be obtained, except in the unlikely event that input proportions on

LPF are identical with those on MPF, or there are compensating variations. Thus the 60 most productive nonfruit farms were 96 per cent more productive with MPF inputs used as weights and 108 per cent more productive with LPF inputs used. As in any index, the results obtained are not independent of the weights employed.

It is apparent from Table 31 that if the various groups of LPF had achieved the same levels of productivity as the MPF, total output would have been substantially higher than it was with no increase in the quantities of measured inputs. For example, given the mean quantities of inputs employed on the 37 LPF in the annual crop group, these farms would have produced an average of $E^O 8,255$ (geometric mean) instead of only $E^O 4,509$, an increase of 83 per cent. While these precise results cannot be safely generalized to the country as a whole, the situation at least in the irrigated portion of the Central Valley – the agricultural heartland of the nation – was probably broadly similar to that in O'Higgins. It was demonstrated in the previous chapter, for example, that the pattern of resource use and returns in Santiago Province and elsewhere in the Central Valley in the early 1960's resembled that in O'Higgins in the late 1950's in some important respects. Hence it is probably not too much to say that, for the country as a whole, if the average level of farm productivity had been as high as that of the top 20 to 25 per cent of farms, the performance of the agricultural sector would have been considerably more impressive than it was. Average farm income would have been much higher, the balance of payments more easily manageable, and the entire economy considerably stronger.

Table 31.   Actual and Hypothetical Production and Productivity Ratios, MPF and LPF

| Type of farm | Average production (E^O) | | Productivity ratios |
| | Actual | Hypothetical | |
	(1)	(2)	(3) = (1) ÷ (2)
Nonfruit (177):			
60 MPF	12,500	6,393	1.96
60 LPF	2,390	4,881	0.48 (2.08)
Livestock (41):			
20 MPF	6,971	4,579	1.52
21 LPF	3,081	4,584	0.67 (1.49)
Annual crops (105):			
37 MPF	11,110	6,080	1.83
37 LPF	4,509	8,255	0.55 (1.83)
Unspecialized (64):			
32 MPF	5,610	3,368	1.67
32 LPF	3,417	5,299	0.64 (1.55)

Note: Calculation of hypothetical production is described in the text. Numbers in parentheses after each group heading are total numbers of farms in the group. Numbers before MPF and LPF are numbers of farms used in calculating MPF and LPF production functions. Numbers in parentheses in column (3) are reciprocals of the LPF productivity ratios. Actual production is the geometric mean for each group.

Sources of Productivity Differences: Prices

It is of considerable interest, therefore, to try to understand why productivity differences were so pronounced. There are several possibilities. One is that, for whatever reason, MPF were in a position to sell their output more dearly than LPF. In this case they would appear to be producing more than LPF per unit of input even if the physical input-output relations were identical.

The data permitted partial exploration of this hypothesis with respect to annual crop farms. Prices received for wheat, corn, potatoes, beans, and sunflower seed (the major crops on both LPF and MPF) were studied. The results are shown in Table 32. They indicate that the MPF received higher prices for all five of the commodities analyzed. The differences were so slight, however, that they could not have been responsible for much of the 83 per cent productivity differences between the more and less productive annual crop farms.

Table 32.   Prices Received, 37 Most Productive and 37 Least Productive Annual Crop Farms, O'Higgins Province, 1958/59

	Prices received (E° per quintal)		
	MPF	LPF	MPF ÷ LPF
Wheat	6.65	6.43	1.034
Corn	5.19	5.09	1.020
Potatoes	3.72	3.42	1.088
Beans	12.80	12.42	1.031
Sunflower seed	9.41	8.57	1.098

Source: Survey 1959.

Sources of Productivity Differences: Quality of Inputs

It appears, therefore, that much the greater part of the observed differences in productivity must be attributable to other factors. It is helpful to begin the analysis of these by measuring the relative contribution of the various inputs to the observed differences in total productivity. Tables 33 through 36 show this for the various groups of LPF and MPF. The actual output of any input in any LPF group was found by multiplying the amount of that input by its own marginal product. The hypothetical output for the same input was found by multiplying it by the marginal product of that input in the corresponding MPF group. For example, the average quantity of land on the less productive annual crop farms (Table 35) was multiplied by its own marginal product to obtain the actual output of land on those farms, namely E°1,246. This same quantity of land was then multiplied by the marginal product of land on more productive annual crop farms to obtain the hypothetical product of this input, namely $E^O1,753$. The difference between the actual and hypothetical product of land was therefore E^O507 (column 3 of Table 35). Since the total productivity difference between the two groups of farms was $E^O3,746$ (based on LPF inputs and MPF marginal products), the contribution of land to this difference was 13.5 per cent (column 4 of Table 35). The estimates in the bottom half of each table were calculated in the same way except that MPF inputs were multiplied by MPF marginal products to obtain actual output and by LPF marginal products to obtain hypothetical output.

One possibly important source of the differences in productivity shown in Tables 33 through 36 is differences in the quality of resources used. The quality differences may exist between inputs represented in the production function or between unmeasured inputs. Climate, management, and water, for example, do not appear in the production functions. As noted earlier, climatological conditions are fairly uniform in the area of the study; differences in management quality, however, may be a major source of productivity differences, and more will be said on this below. Since all the farms in the sample were irrigated, the failure to include water in the production functions would bias the productivity results only if some farms received too little water for optimal irrigation, or if their receipt of it was untimely, or if their land was too hilly or otherwise difficult to irrigate properly. The survey data provide no information on these specific points. However, they do reveal that about 40 per cent of the LPF reported (unspecified) "problems" in the supply of irrigation water, which affected about 12 per cent of their total irrigated area. Among the MPF only 1 per cent of the irrigated area was affected by problems of water supply. This suggests that some of the observed productivity differences resulted from differences in the availability of water. While we cannot precisely estimate the importance of this factor, the small proportion of land on the LPF suffering from water problems suggests that its contribution to the total productivity differences was small.

There may also be quality differences within classes of inputs included in the production functions. Land is a case in point. In computing the production functions for the 60 most productive and least productive nonfruit farms, quality differences in land were eliminated; hence the differences between these two groups in the productivity of land could not reflect land quality differences. Nonetheless, if the 60 MPF had significantly better land than the 60 LPF, this would tend to raise the productivity of *nonland* inputs on the MPF above that on the LPF. The ratio of top quality land to total land was .585 on the MPF and .534 on the LPF, indicating that in fact the MPF had somewhat better land. The difference appears small, however, particularly in proportion to the very large difference in total productivity between the two groups.

When the farms are divided into livestock, annual crop, and unspecialized farms, the differences between MPF and LPF with respect to the ratios of top

Table 33. Contribution of Inputs to Actual and Hypothetical Production, 60 Most Productive and 60 Least Productive Nonfruit Farms

	Production ($E°$)		Col. (2) minus col. (1)	Per cent contribution of each input to total productivity difference[a]
	Actual	Hypothetical		
	(1)	(2)	(3)	(4)
Based on LPF inputs, MPF marginal products:				
Land	708	1,191	483	19.4
Labor	474	1,825	1,351	54.3
All operating expenses except fertilizers, pesticides, and seeds	736	1,021	285	11.4
Seeds	125	273	148	5.9
Fertilizers and pesticides	165	185	20	0.8
Machinery	73	259	186	7.5
Animals	73	99	26	1.0
Productive structures	36	28	-8	-0.3
Total	2,390	4,881	2,491	100.0
Based on MPF inputs, LPF marginal products:				
Land	2,638	1,560	-1,078	17.7
Labor	4,113	1,069	-3,044	49.8
All operating expenses except fertilizers, pesticides, and seeds	3,079	2,223	-856	14.0
Seeds	901	411	-490	8.0
Fertilizers and pesticides	581	517	-64	1.0
Machinery	782	217	-565	9.3
Animals	229	168	-61	1.0
Productive structures	177	228	51	-0.8
Total	12,500	6,393	-6,107	100.0

[a]These percentages were found by dividing each entry in col. (3) by the total of these entries and multiplying by 100.

Table 34. Contribution of Inputs to Actual and Hypothetical Production, 20 Most Productive and 21 Least Productive Livestock Farms

	Production (E°)		Col. (2) minus col. (1)	Per cent contribution of each input to total productivity difference[a]
	Actual	Hypothetical		
	(1)	(2)	(3)	(4)
Based on LPF inputs, MPF marginal products:				
Land	359	345	–14	–0.8
Labor	111	333	222	14.8
Operating capital	1,626	2,344	718	47.8
Livestock	420	808	388	25.8
Machinery and structures	565	754	189	12.6
Total	3,081	4,584	1,503	100.0
Based on MPF inputs, LPF marginal products:				
Land	501	521	20	–0.8
Labor	486	162	–324	13.5
Operating capital	2,913	2,021	–892	37.3
Livestock	1,856	964	–892	37.3
Machinery and structures	1,215	911	–304	12.7
Total	6,971	4,579	–2,392	100.0

[a]These percentages were found by dividing each entry in col. (3) by the total of these entries and multiplying by 100.

Table 35. Contribution of Inputs to Actual and Hypothetical Production, 37 Most Productive and 37 Least Productive Annual Crop Farms

	Production (E°)		Col. (2) minus col. (1)	Per cent contribution of each input to total productivity difference[a]
	Actual	Hypothetical		
	(1)	(2)	(3)	(4)
Based on LPF inputs, MPF marginal products:				
Land	1,246	1,753	507	13.5
Labor	1,408	3,101	1,693	45.3
"Other" operating capital	1,537	2,016	479	12.8
Seed	188	564	376	10.0
Fertilizer	74	486	412	11.0
Machinery	116	338	222	5.9
Livestock and structures	−60	−3	57	1.5
Total	4,509	8,255	3,746	100.0
Based on MPF inputs, LPF marginal products:				
Land	2,145	1,524	−621	12.3
Labor	4,200	1,907	−2,293	45.6
"Other" operating capital	2,899	2,211	−688	13.8
Seed	891	297	−594	11.8
Fertilizer	573	88	−485	9.6
Machinery	461	159	−302	6.0
Livestock and structures	−59	−106	−47	0.9
Total	11,110	6,080	−5,030	100.0

[a]These percentages were found by dividing each entry in col. (3) by the total of these entries and multiplying by 100.

Table 36. Contribution of Inputs to Actual and Hypothetical Production, 32 Most Productive and 32 Least Productive Unspecialized Farms

	Production (E°)		Col. (2) minus col. (1)	Per cent contribution of each input to total productivity difference[a]
	Actual	Hypothetical		
	(1)	(2)	(3)	(4)
Based on LPF inputs, MPF marginal products:				
Land	319	-463	-782	-41.6
Labor	1,366	2,547	1,181	62.8
"Other" operating capital and seed	1,098	1,826	728	38.7
Fertilizer	626	636	10	0.5
Machinery	-330	59	389	20.7
Livestock	295	617	322	17.1
Structures	43	77	34	1.8
Total	3,417	5,299	1,882	100.0
Based on MPF inputs, LPF marginal products:				
Land	-432	298	730	-32.6
Labor	2,581	1,386	-1,195	53.4
"Other" operating capital and seed	2,138	1,288	-850	37.9
Fertilizer	404	397	-7	0.3
Machinery	73	-413	-486	21.7
Livestock	771	369	-402	17.9
Structures	75	43	-32	1.4
Total	5,610	3,368	-2,242	100.0

[a]These percentages were found by dividing each entry in col. (3) by the total of these entries and multiplying by 100.

quality to total land become somewhat greater. They still are small, however, in relation to the differences in total productivity.

The data do not permit investigation of quality differences in fixed capital or in the kinds of pesticides used. This is true also of seed inputs. Some information was collected on the quality of seeds used, but, unfortunately, only two basic categories were distinguished, ordinary ("corriente") and selected ("seleccionada"), and the differences between these categories are vague. A seed variety considered by one farmer to be "corriente" may be considered "seleccionada" by another. This is evident in the responses to questions on this point. There are a large number of cases in which identical prices are given for seeds called "seleccionada" by one farmer and "corriente" by another. Since prices for better quality seed were in fact higher, it is apparent that farmers had differing views regarding the meaning of "seleccionada" and "corriente."

Detailed examination of the data on fertilizer use on annual crop farms showed that virtually the only types employed were nitrates and potassics. The mix of them was about the same on both MPF and LPF. Moreover, there were only a small number of suppliers of fertilizers – the State Bank being the most important – and they were of about equal relative importance as sources of fertilizers to both LPF and MPF. Hence, there is no reason to believe that quality differences in fertilizers contributed anything to the total productivity differences.

In calculating the production functions it was not possible to take account of differences in labor quality. This did not appear a serious limitation, because the great mass of farm workers in Chile in the 1950's had only a minimum of formal education or training and hence would generally be classed as unskilled. Nonetheless, there may have been significant differences in the quality of farm workers. If this range of qualities were not evenly distributed among farms – and it may not have been if better managers sought out better workers – then labor inputs would cause differences in productivity. This is of special importance because Tables 33 through 36 show that, except for livestock farms, labor accounted for about 50 per cent or more of the total productivity differences. Obviously, if the differences in labor productivity can be explained a major step will be made in explaining total productivity differences. It is of considerable interest, therefore, to examine the extent to which labor quality differences may account for the total differences in labor productivity.

If workers were paid according to their contributions to production – which would be the case if the labor market were working properly – then better quality workers would receive higher wages than those of lesser quality. Therefore, if MPF employed better quality labor than LPF we would expect to find that MPF on the average paid higher wages than LPF. Table 37 indicates that in fact more productive livestock and annual crop farms did pay slightly higher wages, but that on unspecialized farms the reverse was

Table 37. Average Daily Wages Paid on Farms in O'Higgins Province, 1958/59

		(Escudos)
Type of farm	MPF	LPF
Livestock	0.856	0.825
Annual crops	0.856	0.822
Unspecialized	0.779	0.799

Source: Survey 1959. Wages include cash payments plus the value of food given in kind, social insurance, and the imputed value of rights to housing and land provided to the resident labor force. The average wage for unskilled workers was imputed to the labor of owners who were directly engaged in the physical processes of production.

true. For the first two groups, therefore, the wage data are consistent with the hypothesis that MPF employed better quality labor than LPF, given the assumption that wage differences reflect quality differences. However, the wage differences are very small in relation to the observed differences in labor productivity. This may be due in part to measurement errors. The average wage of unskilled workers was imputed to owners directly involved in physical labor on the farm. This would be important only in those cases where the owner's labor is the major part of the total. There were a considerable number of such cases, however, and while there were proportionately more in the LPF group than in the MPF group, they were not insignificant in the latter. Hence any differences in labor skills between owners of MPF and LPF would be obscured by the procedure adopted.

Another possibility is that workers were not paid precisely in proportion to skill, or more exactly, that the proportional differences between marginal products and wages were greater for skilled workers than for unskilled workers. This could occur if the demand and supply curves for unskilled labor were considerably more elastic than the corresponding curves for skilled labor, and if for whatever reason the employment of each type was less than optimum. This is illustrated in Figure 1. *WS* is the supply curve for unskilled labor, indicating that it could be hired up to the amount *OB* without any increase in the minimum wage of *OW*. *DCD* is the marginal productivity curve for unskilled labor while *DCA* is its demand curve. *DCA* reflects the assumption that the demand for unskilled labor will be such as to maintain equality between its wage rate and its marginal product up to the point *OA*. No more labor is demanded beyond this point, even though the wage rate *OW* is less than the marginal product *OMP*. Since *OB* of labor is offered at the wage *OW*, there is unemployment of unskilled labor equal to *AB*. This situation is consistent with that already shown to have existed in Chilean agriculture in the 1950's: the great bulk of the labor force unskilled, the marginal product of labor somewhat above the going wage,[10] and some, perhaps much, unemployment. Considerable evidence was presented to show that the funda-

[10]Cf. Table 23.

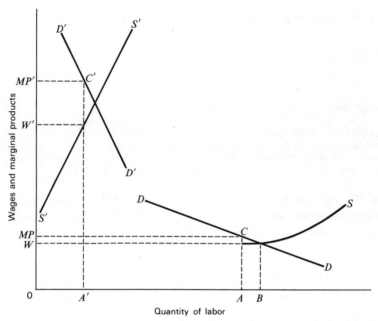

Figure 1. Illustration of situation in which difference between marginal product and wage is proportionately greater for skilled than for unskilled labor.

mental cause of this underemployment of labor was a general shortage of nonland and nonlabor inputs.

The curve $S'S'$ is the supply curve for skilled labor. None of this is forthcoming at the minimum wage OW, and in general the relative response to given percentage increases in wages is lower than for unskilled labor. This reflects the obvious fact that the absolute number of skilled workers is very small and that to increase the supply at any given time is difficult. $D'C'A'$ is the demand curve for skilled labor while $D'C'D'$ is its marginal product curve. The latter is drawn less elastic than DCD on the assumption that the complementarity between skilled labor and machinery, fertilizers, pesticides, improved seeds, and animal stocks is higher than the complementarity between unskilled labor and these inputs. Hence, as the amount of skilled labor increased in relation to them, its marginal product would decline relatively faster than that of unskilled labor.

Given the assumptions, the ratio of the marginal product of skilled labor (OMP') to its wage (OW') would be greater than the corresponding ratio $\left(\dfrac{OMP}{OW}\right)$ for unskilled labor. That is to say, the differences in wages paid to skilled and unskilled labor would not fully reflect differences in productivity. Hence the fact that the wage differences shown in Table 37 are much smaller than the differences in labor productivity does not necessarily lead to rejection of the hypothesis that the MPF employed better quality labor than the LPF.

Data collected on the distribution of labor by skill categories on annual crop farms provide some slight additional support for this hypothesis. These are shown in Table 38. This table indicates that MPF employed proportionately more of both skilled and unskilled labor. However, it is likely that most of the "unspecified" category on LPF were unskilled. Moreover, the number of very small farms was proportionately larger in the LPF group. The owners of these farms are listed in the first category, although in fact many of them would be rated as unskilled. Hence the true proportion of unskilled workers on LPF was almost surely higher than that on MPF.

Table 38. Percentage Distribution of Days Worked by Skill Categories, Most Productive and Least Productive Annual Crop Farms, O'Higgins Province, 1958/59

	MPF	LPF
Owners, administrators, and chief foremen	8.3	13.7
Skilled and semiskilled labor	6.9	1.4
Unskilled:		
Resident laborers	39.5	28.7
Hired laborers	43.0	47.4
Unspecified	2.3	8.8
Total	100.0	100.0

Source: Survey 1959.

Hence both the wage data and the distribution-by-skill data are consistent with the hypothesis that the differences in labor productivity on MPF and LPF in part reflect differences in the quality of the labor force. However, both the wage differences and the skill differences are quite small in relation to the observed differences in labor productivity. To be sure, there may be measurement errors and wages may not fully reflect productivity differences. Any reasonable allowance for these factors, however, would seem to leave a substantial part of the labor productivity differences unaccounted for.

Sources of Productivity Differences: Factor Proportions

All other things being equal, the productivity of any input will be higher the greater the amount of other inputs with which it is combined. If labor on more productive farms were combined with more nonlabor inputs than on less productive farms, this might help to explain the much greater productivity of labor on MPF than on LPF. Tables 39 through 41 present the relevant information for the various nonlivestock farm groups.[11] The tables

---

[11] Data are not presented for livestock farms because labor was of little importance in the total productivity difference between farms in this group. This is consistent with the point made a number of times in this study that the Chilean livestock industry behaves quite differently in some respects from the rest of the agricultural sector. For this reason the analysis of productivity differences will concentrate on the nonlivestock groups of farms.

Table 39.   Amount of Input Per Unit of Land and Labor, 60 Most Productive and 60
Least Productive Nonfruit Farms

| | Input per unit of – | | | |
| | Land | | Labor | |
Inputs	MPF	LPF	MPF	LPF
Land	1.00	1.00	0.015	0.016
Labor	65.5	64.4	1.00	1.00
Operating expenses except seeds, fertilizers,				
and pesticides	50.5	37.2	0.771	0.578
Seeds	14.9	10.0	0.227	0.155
Fertilizers and pesticides	3.1	2.2	0.048	0.034
Machinery	98.9	73.0	1.51	1.13
Livestock	78.8	75.6	1.20	1.17
Productive structures	76.5	27.0	1.17	0.42
Total operating expenses[a]	74.5	56.8	1.14	0.88
Total fixed capital[b]	368.6	257.3	5.63	3.99

Note: The sums of the ratios for the components of total operating expenses and total fixed capital do not equal the ratios for these totals, respectively, because the ratios are calculated from the geometric means of the inputs.

[a]Value of seeds, fertilizers, pesticides, and all other operating expenses.
[b]Value of stock of machinery, livestock, and productive structures.

show the amounts of various inputs per unit of land and labor. Information is included for land as well as labor for two reasons. In the first place, except on unspecialized farms, differences in the productivity of land are important in total productivity differences; hence land merits attention on this score alone. Secondly, land, like unskilled labor, is a "traditional" input, the productivity of which – given its inherent fertility – clearly depends upon the amounts of other inputs combined with it. Studies have shown that countries which have achieved rapid increases in agricultural production and rising incomes for farm people have done so by greatly increasing the amounts of nonland, nonlabor inputs in relation to the quantities of land and labor employed.[12] This suggests that when groups of farms in any given period of time are compared, those with larger amounts of nonland and nonlabor inputs per unit of land and labor will show higher returns to land and labor than farms for which these amounts are less.

Tables 39 to 41 in general show that more productive farms employed more nonland, nonlabor inputs per unit of land and labor than less productive

[12]Studies of the American and Japanese experience present quite unequivocal evidence on this score. Data from two such studies are given in Bruce F. Johnston and G. S. Tolley, "Strategy for Agriculture in Development," *Journal of Farm Economics*, 47 (1965), 367. The studies referred to are Saburo Yamada, "Long-Term Changes in Agricultural Inputs and Output" (Tokyo: 1963), and R. A. Loomis and G. T. Barton, *Productivity of Agriculture, United States, 1870-1959*, U. S. Department of Agriculture, Technical Bulletin No. 1238.

Table 40. Amount of Input Per Unit of Land and Labor, 37 Most Productive and 37 Least Productive Annual Crop Farms

| | Input per unit of − | | | |
| | Land | | Labor | |
Inputs	MPF	LPF	MPF	LPF
Land	1.00	1.00	0.027	0.030
Labor	37.3	33.7	1.00	1.00
Operating expenses except seeds, fertilizers, and pesticides	25.4	21.5	0.681	0.638
Seeds	9.21	7.08	0.241	0.211
Fertilizers and pesticides	2.01	2.09	0.054	0.062
Machinery	50.4	45.1	1.35	1.34
Productive structures	45.5	22.2	1.22	0.659
Total operating expenses[a]	39.6	35.3	1.07	1.06
Total fixed capital[b]	178.1	128.7	4.81	3.86

Note: The sums of the ratios for the components of total operating expenses and total fixed capital do not equal the ratios for these totals, respectively, because the ratios are calculated from the geometric means of the inputs.

[a]Value of seeds, fertilizers, pesticides, and all other operating expenses.
[b]Value of stock of machinery, livestock, and productive structures.

Table 41. Amount of Input per Unit of Land and Labor, 32 Most Productive and 32 Least Productive Unspecialized Farms

| | Input per unit of − | | | |
| | Land | | Labor | |
Inputs	MPF	LPF	MPF	LPF
Land	1.00	1.00	0.029	0.032
Labor	34.5	31.3	1.00	1.00
Operating expenses except seeds, fertilizers, and pesticides	29.0	20.6	0.841	0.658
Seeds	5.17	6.33	0.150	0.202
Fertilizers and pesticides	0.931	1.36	0.027	0.043
Machinery	47.2	35.1	1.37	1.11
Livestock	54.5	40.5	1.58	1.28
Productive structures	15.9	15.2	0.461	0.486
Total operating expenses[a]	38.3	31.9	1.11	1.02
Total fixed capital[b]	175.2	130.6	5.08	4.18

Note: The sums of the ratios for the components of total operating expenses and total fixed capital do not equal the ratios for these totals, respectively, because the ratios are calculated from the geometric means of the inputs.

[a]Value of seeds, fertilizers, pesticides, and all other operating expenses.
[b]Value of stock of machinery, livestock, and productive structures.

farms. This is true without exception in the case of the 60 MPF and LPF selected from the group of 177 nonfruit farms (Table 39). The man-land ratios on those two groups of farms were about the same, but the 60 MPF used more of every input per unit of land and labor than the LPF. While the pattern is somewhat less clear for the annual crop and unspecialized farms, it is still unmistakable. In both cases the amounts of operating and fixed capital per unit of land and labor were greater on MPF than on LPF. All three tables, therefore, are consistent with the hypothesis that the productivity advantages of land and labor on MPF were owed to the greater amounts of other inputs with which they were combined.[13]

The skeptical reader may question this conclusion on the grounds that in some cases the differences in the amounts of nonland, nonlabor inputs per unit of land and labor are quite small. This is particularly true in the case of labor per unit of machinery on annual crop farms, where there is virtually no difference. This is of special importance because the productivity of labor is likely to be particularly sensitive to the amount of machinery with which each worker is equipped. If these amounts are the same on two groups of farms, what grounds are there for believing that machinery inputs have any impact on differences in labor productivity?

While there is no sure response to this question, several possible explanations come to mind. One is that there were differences in the quality of machinery employed on the two groups of farms, differences not adequately reflected in machinery values. As noted above, it was not possible to distinguish machinery qualities in the sample data, so this alternative cannot be explored here.

A second possible explanation can be pursued at some length, however. It will be recalled that machinery (and other fixed capital) inputs were measured by the *stock* of machinery in each farmer's inventory at the end of the crop year. If more productive farmers achieved greater utilization of their machinery over the course of the year than less productive farmers, the *actual* input of machinery services per unit of labor on MPF would be greater than on LPF even though the ratios of machinery stock to labor were the same. It is, of course, the amount of machinery services actually used per unit of labor which is important in explaining differences in labor productivity.

There is good reason to believe that MPF achieved fuller utilization of their stock of farm machinery than LPF. The rate of machine utilization is

---

[13] This statement clearly does not apply to land productivity differences on unspecialized farms, since in this group land was less productive on MPF than on LPF. This is an apparent exception to the argument made in the text. The problem may lie in poor statistical estimates of land productivity on MPF and LPF in this group. Table 30 shows that neither of these estimates was statistically different from zero, a result attributable perhaps to the small size of the two farm groups (32 in each) and the high intercorrelation between land and some other inputs. Moreover, land in these groups was not adjusted for quality differences. Had it been possible to do this and had the sample sizes been larger, the results might have been different.

particularly sensitive to size of farm, larger farms typically achieving better utilization than smaller ones. This is because farm machinery comes in bulky units, the minimum size of which will be too large for full utilization on farms below some given size. The minimum size tractor and plow combination, for example, may be sufficient to plow 5 acres in one day. A 5-acre farm, therefore, would employ such a tractor only a few days per year, while utilization on a 100-acre farm would be substantially greater. If the man-land ratios on the two farms were about the same, then the amount of machine services per unit of labor would be substantially greater on the larger farm, with a corresponding favorable impact on labor productivity.

Sources of Productivity Differences: Size of Farms

The question of the average size of MPF and LPF thus presents itself. Table 42 indicates that except for the unspecialized farms, MPF were larger than LPF, when size is measured by number of irrigated hectares.[14] Although the differences were statistically significant at conventional probability levels only for the 60 most and least productive nonfruit farms, they nonetheless support the hypotheses that MPF achieved better utilization of farm machinery than LPF (with the exception, of course, of the unspecialized farms).

[14]Where the major farm inputs are land and labor, and they are used in more or less fixed proportions, land area is quite an adequate measure of farm size. However, where other inputs account for a substantial share of production, it is useful to take them into account when measuring farm size. This can be done by means of indexes of input quantities. Such indexes can be constructed by dividing the actual output of one group of farms by the hypothetical output of another group of farms. The result is an input index weighted by the marginal products on the first group of farms. A similar index, weighted by marginal products of the second group of farms, can be constructed by dividing hypothetical output of the first group of farms by actual output of the second group.

The following table shows input indexes for MPF and LPF:

Indexes of Total Input (LPF = 100)

	Weighted by –	
	MPF marginal products	LPF marginal products
Nonfruit farms	256	267
Livestock farms	152	149
Annual crop farms	135	135
Unspecialized farms	106	99

These indexes were constructed in the manner described above, the actual and hypothetical outputs being taken from Table 31. The indexes show that, except for the unspecialized farms, the MPF were substantially larger than the LPF. Moreover, the differences are more pronounced when size is measured by total inputs than when it is measured by area alone (see Table 42). This indicates that while MPF had considerably more land than LPF, their supplies of nonland inputs were proportionately even greater.

Table 42. Average Size of MPF and LPF, Measured by Irrigated Area

| | Average area (hectares) | | | |
| | Arithmetic means | | Geometric means | |
Type of farm	MPF	LPF	MPF	LPF
Nonfruit	118.1	70.8	62.5	30.9
Livestock	114.2	66.1	39.9	27.5
Annual crops	97.8	75.0	57.5	47.0
Unspecialized	80.6	114.7	39.9	42.8

Source: Size by hectare from Survey 1959.

Note: The differences between arithmetic means were tested to determine if they were statistically significant, with the following results:

	Probability of difference as large as that observed if true difference were zero
Nonfruit farms	2%
Livestock farms	15%
Annual crop farms	26%
Unspecialized farms	30%

This hypothesis is of special relevance to the annual crop farms because it was in this group that the stock of machinery per man was the same for MPF and LPF. It is particularly important, therefore, to determine if the larger size of the most productive annual crop farms permitted them to achieve fuller utilization of machinery than was possible for the least productive farms. The first step was to take a closer look at the relation between size and productivity on annual crop farms. These farms were grouped into various size categories and average productivity ratios calculated for each category. The results are shown in Table 43. They show no clear trend in the relation between size and productivity for farms from 1 to 10 hectares up to 100.1 to 150 hectares. Above 150 hectares, however, productivity shows a very sharp increase.[15]

If the sudden jump in productivity on farms above 150 hectares was due to fuller utilization of farm machinery, then independent evidence ought to show that (1) the minimum size for efficient use of machinery was around 150 hectares, and that (2) machinery rental services were not available or were expensive. The second condition is necessary because machinery rental is potentially a substitute for machinery ownership. Even though operators of farms under 150 hectares may not have been able to justify the purchase of farm machinery, they still could have obtained the same mix of machinery

[15]The significance of the differences between productivity ratios by size was tested by variance analysis. None of the differences were significant at the 1 per cent level of probability. The differences between farms from 1 to 50 hectares and those from 150.1 to 200 and from 200.1 to 250 hectares were significant, however, at the 5 per cent level.

Table 43.   Relation between Farm Size and Productivity Ratios, Annual Crop Farms,
O'Higgins Province, 1958/59

Size (hectares)	Average productivity ratios[a]
1.0–10.0	0.818 (12)
10.1–20.0	0.762 (11)
20.1–30.0	0.860 (11)[b]
	0.963 (12)[b]
30.1–40.0	0.769 (10)
40.1–50.0	0.945 (10)
1.0–50.0	0.852 (55)
50.1–100.0	0.919 (18)
100.1–150.0	0.812 (5)
150.1–200.0	1.089 (11)
200.1–250.0	0.961 (8)[c]
	1.095 (9)[c]
250.1+	0.941 (7)

[a]The ratio for any farm is its actual output divided by its hypothetical output. Average ratios are unweighted. Figures in parentheses are the number of farms in each size category.
[b]One farm in this group had a productivity ratio of 2.098. When this farm is excluded the average for the remaining 11 is 0.860.
[c]One farm in this group had a productivity ratio of 2.164. When this farm is excluded the average for the remaining 8 is 0.961.

and labor services as farmers over 150 hectares if machinery rentals were available at economical prices.

In the absence of published studies dealing with these questions, two officials in the Corporación de Fomento (CORFO) were consulted independently. Both were specialists in the analysis of the country's needs for farm machinery. One indicated that a farm in O'Higgins Province specializing in annual crops would have to have a minimum of 125 irrigated hectares to justify economically the purchase of a tractor with attendant equipment. The other estimated that 200 irrigated hectares is the minimum for farms in that area.

These estimates are consistent with the hypothesis that size economies in the use of machinery become important at around 150 hectares. While they reflect the relative prices of machinery and farm products of 1966/67, these apparently were not markedly different from those of 1958/59.[16] It is interesting to note also that the CORFO estimates are generally consistent with what is known about the relationship between farm size and machine economies in the United States. Studies done in 1960 of family farms in southern Iowa specializing in annual crops showed that minimum costs with a

[16]*El uso de maquinaria* (see Ch. 2, n. 4), p. 30.

two-plow or a three-plow tractor were achieved at sizes ranging from 65 to 160 hectares (160 to 400 acres).[17] Since the ratio of farm machinery costs to labor costs was much higher in Chile at that time than in the United States, Chilean farms would have to be larger than those in the United States to justify the purchase of a tractor. Thus the differences between Chile and the United States with respect to size-machine economy relationships were generally consistent with the differing economic conditions confronting farmers in the two countries.

The CORFO experts were also questioned about the availability of machine rental services in O'Higgins Province. They had no information for 1958/1959, but in 1959/1960 SEAM (Servicio de Equipos Agrícolas Mecanizados), an arm of CORFO, provided 37,400 hours of farm machinery rental time to farmers in the province. Data on rental services provided by individual farmers or commercial agencies are not available, although the latter, at least, were not active. About 19,000 hours of the SEAM time was for tractors. This was equivalent to not more than 25 tractors.[18] The 1955 agricultural census shows that in that year there were not fewer than 1,160 tractors in O'Higgins Province.[19] Assuming that this number was not much lower in 1958/1959 — and it was probably higher — the tractor rental services made available by SEAM probably accounted for not more than 2 to 3 per cent of total tractor hours worked in the province.

The agricultural census contains other information indicating that in 1955, 9 per cent of the farms in O'Higgins under 100 hectares "used" tractors and 12 per cent of them rented some type of farm machinery.[20] In contrast, 81

---

[17] J. Patrick Madden, *Economies of Size in Farming*, Agricultural Economic Report No. 107 (Washington: U.S. Department of Agriculture, Economic Research Service, Feb. 1967), pp. 37–39.

[18] *El uso de maquinaria*, p. 69, indicates that all SEAM tractors worked an average of 824 hours each in 1958/59. Data for 1959/60 are not available, but between 1955/56 and 1962/63 tractors averaged 830 hours per year.

[19] Servicio Nacional de Estadística y Censos, *III Censo Nacional Agrícole Ganadero*, II: I, Table 89, p. 211.

[20] These figures are heavily weighted by farms of from 1 to 5 hectares. As farm size increased up to 100 hectares, the percentage that "used" tractors or rented machinery increased, as the following figures show:

O'Higgins Province, 1955

		Percentage of farms which —	
Farm size (hectares)	Number of farms	"Used" tractors	Rented machinery
1–4.9	2,072	2.7	5.9
5–9.9	637	7.4	11.5
10–19.9	457	14.4	21.4
20–49.9	353	30.9	28.9
50–99.9	128	53.1	30.5

Ibid., Table 86, p. 209.

per cent of the farms between 100 and 500 hectares "used" tractors while 45 per cent rented some kind of farm machinery.[21]

These data suggest very strongly that the great majority of farms under 150 hectares neither owned farm machinery nor rented its services. This finding, in combination with that indicating that the minimum size farm for efficient use of machinery is somewhere between 125 and 200 hectares, gives considerable support to the hypothesis that the sharp increase in productivity of farms over 150 hectares reflects economies in the use of machinery.

This hypothesis finds additional backing in the analysis of production functions for larger and smaller farms. Such functions were computed for three size categories of nonfruit farms: (1) 1–50 hectares, (2) 100.1–200 hectares, (3) more than 200 hectares.[22] Some information derived from these functions is presented in Table 44.[23] The table shows very pronounced in-

[21] Ibid. The precise meaning of "used" is not given.

[22] The analysis focused on nonfruit farms rather than annual crop farms because the numbers of the latter in the various size categories were insufficient. This was true also of the number of nonfruit farms in the 150.1–200 hectares category. For this reason we were forced to work with farms of 100.1–200 hectares.

[23] The functions were of the Cobb–Douglas type, calculated on the logarithms of the inputs. They were as follows:

1–50 hectares

$$\log Y = .254 \log X_2 + .218 \log X_3 + .331 \log X_4$$
$$(.096) \qquad (.094) \qquad (.065)$$

$$+ .118 \log X_5 + .074 \log X_6 + .018 \log X_7 - .018 \log X_8$$
$$(.052) \qquad (.030) \qquad (.042) \qquad (.034)$$

$$+ .029 \log X_9 + .263$$
$$(.022)$$

100.1–200 hectares

$$\log Y = .407 \log X_2 + .258 \log X_3 + .461 \log X_4 + .112 \log X_5$$
$$(.155) \qquad (.116) \qquad (.085) \qquad (.095)$$

$$+ .045 \log X_6 + .243 \log X_7 - .067 \log X_8 - .002 \log X_9$$
$$(.034) \qquad (.068) \qquad (.067) \qquad (.016)$$

$$- .672$$

More than 200 hectares

$$\log Y = .414 \log X_2 + .204 \log X_3 + .217 \log X_4 - .001 \log X_5 + .066 \log X_6$$
$$(.247) \qquad (.173) \qquad (.120) \qquad (.018) \qquad (.040)$$

$$+ .096 \log X_7 + .050 \log X_8 - .017 \log X_9 + .172$$
$$(.082) \qquad (.075) \qquad (.104)$$

Where Y is output, X_2 is land adjusted for quality, X_3 is labor, X_4 is all operating expenses except seed, fertilizers, and pesticides, X_5 is seed, X_6 is fertilizers and pesticides, X_7 is machinery, X_8 is livestock, X_9 is productive structures, the final term in each equation is a constant, and the numbers in parentheses are the standard errors of the regression coefficients.

Table 44. Information Derived from Production Functions for Nonfruit Farms of from 1.0 to 50.0 Hectares, 100.1 to 200.0 Hectares, and More than 200.0 Hectares

	Size categories (hectares)		
	1.0–50.0	100.1–200.0	More than 200.0
Sum of regression coefficients	1.024	1.457	1.029
Marginal value products of inputs ($E°$):			
Land	51	117	117
Labor	0.58	1.40	1.12
Operating expenses other than seed, fertilizers, and pesticides	1.73	2.04	0.96
Seed	2.15	1.97	−0.02
Fertilizers and pesticides	7.39	2.88	3.93
Machinery	0.06	0.49	0.26
Livestock	−0.05	−0.16	0.17
Productive structures	0.18	−0.01	−0.04

creasing returns to scale for farms of from 100.1 to 200 hectares, while the other size groups show approximately constant returns.[24] Since the sum of the regression coefficients in these functions is the equivalent of the slope of the regression line in a two variable equation, the pattern in the sums of the coefficients suggests that the relation between farm size and production approximated an S curve. In a two-dimensional diagram in which land represents an index of all inputs the relation could be depicted as shown in Figure 2.

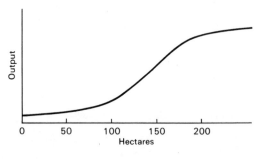

Figure 2.

In addition to the evidence of sharply increasing returns to scale on the 100.1 to 200 hectare farms, the marginal product of machinery in this group was far greater than in either of the other two groups. Indeed it was considerably higher than in any of the various farm groupings that have been considered heretofore. Finally, in each of the larger size groups the marginal products of both land and labor were substantially greater than they were in the 1 to 50 hectare farms.

[24] The sum of the regression coefficients for the 100.1 to 200 hectare group was significantly greater than 1 whereas this sum for the other size groups was not.

There is, therefore, considerable evidence that productivity was positively related to farm size, that the relationship approximated an S curve, and that the inflection points on the curve were somewhere between 100 and 200 hectares. The evidence also suggests that the sharp increase in productivity enjoyed by farms in this size range was due to economies in the use of farm machinery. This evidence is found not only in the production functions for farm-size groups but also in the opinions of Chilean government officials whose business it is to study the economics of farm machinery in the country.

There is another, more indirect way in which size may be positively related to farm productivity, and to the productivity of labor and land in particular. It was noted earlier that MPF used more of *all* inputs relative to land and labor than LPF used. This suggests that the greater productivity of these "traditional" inputs on MPF was not due simply to the greater availability of machinery services, but that it owed something also to a generally more ample supply of "nontraditional" inputs. Farm size may have made this possible by providing larger farms with readier access to credit than smaller farms. If this indirect relation between size and productivity existed then it must be demonstrable that (1) more productive farmers in fact used more credit than less productive farmers and that (2) had less productive farmers used more credit the productivity differences between them and the more productive farmers would have been less.

Unfortunately the sample data do not contain information on the amounts of credit used. However, there is an indication of whether or not credit was used at all, and this information has been examined for each of the 37 most and least productive annual crop farms. Of the 74 farms, 67 answered the question concerning credit, 34 of them among the MPF and 33 among the LPF. Of the answering MPF, 21 farms (63 per cent) used credit and 13 (37 per cent) did not. Among the answering LPF, 14 farms (43 per cent) used credit and 19 (57 per cent) did not.

Thus the first condition specified above seems reasonably well satisfied, at least in the case of annual crop farms. That is, the MPF in this group were larger than the LPF, and they made relatively greater use of credit. In this indirect way, therefore, the larger size of the MPF may have contributed to their greater productivity.

With respect to the second condition — that if LPF had had greater access to credit the productivity differences between them and MPF would have been less — the evidence is more ambiguous. It has already been argued that because of their smaller size LPF could not have made as effective use of farm machinery as MPF. Hence, even if machinery credits had been more readily available to LPF productivity differences would have remained.

The evidence indicates that except for unspecialized farms this was true also of other inputs financed with credit, such as fertilizers, pesticides, and other purchased inputs. Tables 27, 28, and 29 indicate that the marginal products of most of these inputs on LPF not only were less than on MPF but were well under their opportunity costs. The only exceptions were fertilizers and pesticides on the 60 least productive nonfruit farms and all operating expenses except seeds, fertilizers, and pesticides on the 37 least productive annual crop farms. In the latter case the marginal product was $E^O1.66$, close enough to the assumed opportunity cost of $E^O1.50$ to suggest that this group of LPF had little incentive to purchase more of these inputs even if more credit had been available to them.

The 60 least productive nonfruit farms could have profitably used considerably more fertilizers since the marginal product of that input was $E^O4.50$. However, even if they had increased their fertilizer consumption by enough to equalize its marginal product with its cost, the productivity of other inputs still would have remained well below the levels achieved on the MPF. Hence the total productivity of the LPF still would have been well under that of the MPF.

For unspecialized farms the situation may have been different. Both MPF and LPF in this group had enormous marginal returns to expenditures on fertilizers and pesticides. However, the elasticity of production with respect to these inputs was much greater on LPF than on MPF (19.2 per cent and 7.3 per cent respectively). This means that, for given percentage increases in expenditures for fertilizers and pesticides, total output would have increased much faster on LPF than on MPF. The marginal products of fertilizers would have fallen faster on MPF and the marginal products of all other inputs would have risen faster on LPF. Hence, both LPF and MPF could have spent substantially more on fertilizers, but had they done so, the productivity differences between them would have diminished. For example, if LPF had spent five times as much on fertilizers as they actually did, and if MPF had spent 3 times as much as they did, the marginal returns to this input would have been $E^O4.03$ on LPF and $E^O4.27$ on MPF. Thus the returns to fertilizer would still have been approximately equal, but at a much lower level. However, the differences between MPF and LPF in the returns to other inputs would have been much smaller.[25] Hence, the total productivity differences between them would have been reduced. Under the assumptions made, MPF would have been 8 per cent more productive, calculating the productivity index with MPF inputs as weights. Weighting the index by LPF inputs, MPF would have been 9 per cent more productive. With the amounts of fertilizers actually used, the corresponding productivity differences were 67 per cent and 55 per cent.[26]

[25] Since machinery had a negative return on LPF, the difference between returns to this input would have been greater had both MPF and LPF used more fertilizers. As noted in Chapter 4, it is a characteristic of Cobb–Douglas production functions that the signs of marginal products do not change as greater or lesser amounts of inputs are used. In fact, it is likely that had LPF used the assumed larger amount of fertilizers, the marginal product of machinery would have been less negative, or perhaps even positive.

[26] See Table 31.

The illustration suggests that in the unspecialized group of farms, LPF could have substantially narrowed the productivity gap had they employed more fertilizers, even if the MPF also had used more of this input. Whether credit limitations were responsible for the failure of LPF to buy more fertilizer cannot be determined. However, since they were smaller on the average than MPF, there is a presumption that credit at least was not as readily available to them as it was to MPF.

It is unlikely that the apparent inability of LPF in the livestock and annual crop groups to make economical use of more fertilizers, pesticides, and seeds was due to their relatively small size. The mean size of less productive livestock farms was 66.1 hectares whereas for LPF specializing in annual crops it was 75.0 hectares.[27] Fertilizers and pesticides require mechanical devices for most efficient application, but the size of farm necessary to justify investment in these devices is surely not more than 25 hectares.[28]

It has already been indicated that there were no apparent quality differences in the kinds of fertilizers used on LPF and MPF. Because of inadequacies in the data the possibility of quality differences in pesticides and seeds cannot be ruled out; however, existence of such differences is not obvious. If neither size nor quality differences explain the low returns to these inputs on LPF, what does? The most plausible candidate is the input which so far has been left completely out of account: management. The efficient use of fertilizers, pesticides, and seeds requires skill and knowledge of proper dosages, given the qualities of soil, the availability of water, and the kinds of crops cultivated. There is every reason to believe that farmers will vary in the extent to which they possess these skills. Hence when observed differences in the returns to fertilizers and other inputs cannot be explained by size or quality differences, variations in management quality would seem to be the only alternative explanation.

Of course, management differences may explain far more than simply differences in returns to fertilizers, pesticides, and seeds. Indeed, some writers have attributed *all* productivity differences to management, almost, it seems, as a matter of course. Hoch, for example, *defines* the difference between total outputs obtained from a given bundle of inputs as a measure of the difference in management quality.[29] If there were no differences in all the other factors

---

[27] See Table 42.

[28] Johnston and Tolley make the even stronger assertion that the productivity of fertilizer and other "so-called biological-chemical" inputs is neutral with respect to farm size. Bruce F. Johnston and G. S. Tolley, "Strategy for Agriculture in Development," *Journal of Farm Economics*, 47 (1965), 365–83.

[29] Irving Hoch, "Empirical Measurement of Management," in *Farm Management in the West: Problems in Measuring Management*, Report No. 5: *Research into Managerial Ability and Services*, Conference Proceedings, Farm Management Research Committee of the Western Agricultural Economics Research Council, Berkeley, Calif., Jan. 13–15, 1964.

affecting productivity – input and output prices, qualities of observed inputs, climate, size of farm – then by a process of elimination, management would be the only explanatory variable. It could even be argued that differences in input and output prices, input qualities, and size of farm all basically reflect differences in management quality; that is, better managers may be more skillful in buying low and selling high, in selecting and using better quality inputs, and in acquiring land and other assets.

While management may be the only source, or at least by far the most important single source, of productivity differences, it makes little sense simply to assume this to be the case until alternative explanations have been considered. A major problem is that there is no single index of managerial ability. One could draw up a list of the various decisions that farm managers must make and the various operations they must perform, weighting each of these according to some standard of relative importance, and then scoring each farmer according to his performance in each task. The trouble with such a procedure is that both the weighting and the scoring are almost inevitably highly subjective.

No such procedure was attempted in this study. To the extent possible, however, the data for annual crop farms were examined to determine if managerial differences could be detected. Presumably better managers will make wiser choices concerning the kinds of crops to grow and the timing of sowing and harvesting. In our sample, however, there was no discernible difference between MPF and LPF with respect to these activities.

A related question concerns the pattern of land use, that is, the percentage of farm area devoted to various activities. Table 45 shows these percentages for the 37 most productive and 37 least productive annual crop farms. Considering land in production, the only major difference is in the proportion of land devoted to natural pasture. This is a relatively low productivity activity and, all other things being equal, better managers would devote a smaller proportion of their land to this use than poorer managers. However, "all other things" were not equal for MPF and LPF. As noted earlier, about 12 per cent of the LPF irrigated area was affected by problems in the supply of irrigation water, while only 1 per cent of the MPF land was so affected. This, therefore, rather than management quality differences, may explain why LPF had a greater proportion of land in natural pasture than MPF.

More productive farmers averaged 48.4 years of age and 21.7 years of experience as farmers. The less productive farmers were 47.6 years old on the average and had 20.1 years experience as farmers. These differences surely are insignificant. Unfortunately there were no data on farmer education.

Farmers who keep accounts of costs and returns of various farm operations are probably better managers than those who do not. In our sample, 47 per cent of the more productive annual crop farmers kept some record of farm operations while only 25 per cent of the less productive did so.

Table 45.   Percentages of Farm Area Devoted to Various Uses, Most Productive and
            Least Productive Annual Crop Farms, O'Higgins Province, 1958/59

	MPF	LPF
Annual crops	43.3%	44.6%
Fruits, vineyards, and forests	3.3	5.7
Pasture:	25.4	27.4
Natural	8.6	12.8
Cultivated	16.8	14.6
Allocated to resident laborers	5.3	4.6
Indirectly productive	5.2	4.0
Fallow, mountainous, or otherwise unproductive	17.5	13.6
Percentage irrigated area of all farm area	76.0	81.0

Source: Survey 1959.

This last is the only independent piece of information we have suggesting
that MPF were better managed than LPF. While far from conclusive, it
accords with the presumption that at least some of the observed differences
in productivity are attributable to management. We are prepared to accept
this as the principal explanation of the differences in rates of return to
fertilizers, pesticides, seeds, and other operating capital, particularly since
there is no other explanation apparent. However, as already noted, a good
case can be made for the belief that the differences in labor productivity are
attributable to the greater use of machinery and other fixed capital on MPF
than on LPF. The principal evidence supporting this belief is the observation
that productivity on annual crop farms of 150 to 200 hectares is considerably
higher than on smaller farms, that farms of from 100 to 200 hectares exhibit
sharply increasing returns to scale and considerably higher returns to
machinery than other size groups, and that the minimum size of farm
consistent with efficient use of a tractor and attendant equipment is
somewhere between 125 and 200 hectares.

    If better management rather than more machinery and other fixed capital
per man were the explanation of the labor productivity differences, the
implication would be that farmers with 150 to 200 hectares were significantly
better managers than those in the 1 to 50 hectare, 50 to 100 hectare, and 100
to 150 hectare size categories. While one might readily accept the idea that
managers in the smallest group are not as good as those in the 150 to 200
hectare category, it is difficult to see why the latter should be so much better
than farmers with from 50 to 150 hectares. As noted above the information
for annual crop farms in our sample showed no marked differences between
MPF and LPF with respect to cropping and land use patterns or in the timing
of sowing and harvesting. The only positive evidence of better management
on MPF was the greater percentage of farmers that kept accounts.

    On the other hand, there is considerable positive evidence in support of
the hypothesis that the differences in labor productivity were due primarily
to more capital per man on MPF. Without denying that management too may

have played a role, we accept this hypothesis as the one most consistent with the available evidence.

The discussion of the causes of the differences in productivity between MPF and LPF can be summarized as follows:

1. There is evidence that MPF received somewhat higher prices for their products than LPF. However, these differences were not large enough to explain more than a small part of the observed differences in total productivity.

2. A larger proportion of LPF land than MPF land was affected by problems of water supply. This may have contributed in a small way to the differences in total productivity.

3. Differences in the quality of land explain little if any of the productivity differences.

4. There were no significant differences in the quality of fertilizers applied on MPF and LPF.

5. The data do not permit conclusions concerning possible differences in the quality of seeds, pesticides, or farm machinery.

6. For the group of farms taken without regard to specialization as well as for two of the three specialization groups, differences in labor productivity accounted for about one-half of total productivity differences. The MPF may have employed somewhat better quality labor than LPF, but the difference was too small to explain very much of the total labor productivity differences.

7. The MPF in each group employed substantially more fixed capital per unit of labor and per unit of land than the LPF. This was generally true also for each of the major components of fixed capital.

8. The MPF in each group also employed more total operating capital per man and per hectare than the LPF.

9. Except in the unspecialized group, MPF were substantially larger than LPF, whether size is measured by land area or by an index of total inputs. However, the size differences (measured by area) were not statistically significant except for the 60 most and least productive nonfruit farms.

10. To study the relation between size and productivity more closely, annual crop farms were divided into size groups and the average productivity ratios calculated for each group. There was no clear tendency for productivity to rise with size up to 150 hectares. However, productivity for farms of from 150 to 200 hectares was considerably higher than for any other size group. The most pronounced differences were between farms from 1 to 50 hectares and those from 150 to 200 hectares. These differences were statistically significant.

11. Independent evidence indicates that given the soils and typical cropping patterns in O'Higgins Province, the minimum size farm consistent with efficient use of a basic complement of machinery was 125 to 200

hectares. In addition, the supply of farm machinery rental services in O'Higgins in the 1950's was very small. There is strong evidence, therefore, that a major part of the total productivity difference on annual crop farms was attributable to economies in the use of machinery made possible by size of farm. This conclusion is strengthened by the finding that farms of from 100 to 200 hectares exhibited strongly increasing returns to scale, while those in other size classes did not.

12. This conclusion is consistent with the finding that the single most important productivity difference was in labor. By virtue of size, larger farms were able to employ efficiently more machinery and other fixed capital per man than smaller farms. The effect was to raise the marginal product of labor on larger farms well above that on smaller farms.

13. More productive farms made greater use of credit than less productive farms. For unspecialized farms the productivity difference between MPF and LPF would probably have been much smaller had the LPF used relatively more fertilizers and pesticides. Whether they failed to do so because of inadequate credit cannot be determined.

14. For the other farm groups, inadequate credit cannot explain the productivity differences with respect to fertilizers, pesticides, seeds, and other items of operating capital. The returns to these inputs on LPF were generally below their costs. Hence even if more credit had been available to LPF to purchase these inputs, they could not have used it effectively.

15. Limited size of farm rather than inadequate credit was the major reason why LPF in the various groups did not employ more machinery and other fixed capital per man.

16. It is unlikely that size of farm explains the low returns to the various items of operating capital on LPF, since the minimum size consistent with efficient use of these inputs must be considerably less than average size of LPF.

17. The differences in returns to these inputs were probably attributable to management quality differences. There is little positive evidence in support of this conclusion, however. It is reached primarily by a process of elimination of alternative explanations.

18. Farmers who are better in the choice and application of fertilizers, pesticides, and so on may also be better in all other aspects of farm operation. Conceivably, therefore, management quality differences could explain the labor productivity differences as well as those for operating capital. To accept this hypothesis, however, would imply that managers on farms of 150 to 200 hectares were substantially better than those on farms of 50 to 150 hectares. There is no obvious reason to expect this. The more-fixed-capital-per-man hypothesis, on the other hand, is plausible and consistent with the available evidence. This, rather than the management-quality hypothesis, provides the more convincing explanation of the observed differences in labor productivity.

CHAPTER

6

Summary of Conclusions and Implications for Agricultural Planning

SUMMARY

The purpose of this study has been to investigate the performance of Chilean agriculture in the 1950's to determine wherein it fell short of its potential and to offer some tentative explanations for the shortfall. In effect, two measures of performance were used: One compared the growth of production generally against some rough measure of what might have been achieved; the second compared the productivity of groups of more productive and less productive farms, the difference providing an indication of production which could have been achieved if the productivity of the low group could have been brought up to the level of the high group.

The major conclusions concerning performance judged by these two measures have already been noted in Chapters 4 and 5 and will not be restated in full detail here. Some general statements about these conclusions are worth noting, however. On the whole, those concerning the growth of production and patterns of resource use in the 1950's are more firmly grounded than the conclusions regarding productivity differences. This is because it was possible to test some of the more important of the former against a number of independent sets of data, while the conclusions concerning productivity are drawn solely from analysis of the 210 farms in O'Higgins Province. One of the major conclusions about the growth of output and patterns of resource use was that farmers at the end of the 1950's could have profitably employed considerably more of all nonland resources, except livestock. This conclusion in turn was based on the finding that rates of return to irrigated land were well below the farmer's opportunity cost of capital while the returns to fertilizer and pesticides were substantially above their opportunity costs. This pattern of returns was found among the 177 nonfruit farms in O'Higgins in 1958/1959 and among three other independently sampled sets of farms in

Santiago, O'Higgins, and other provinces in the Central Valley in the mid-1950's and early 1960's. Moreover, an analysis of national data for the 1950's as a whole yielded estimates of the return to fertilizers and pesticides which were quite consistent with those based on provincial data.

The fact that these various sets of independently derived data all reveal the same pattern of returns to land and fertilizer strongly suggests that the pattern was not a statistical mirage but a genuine fact of life in Chilean agriculture. This in turn gives considerably more credence to the conclusion that more nonland inputs could have been profitably employed than would be the case if the various data sets were not so consistent with one another or if only a single set were available.

The conclusion that the failure to seize the production opportunities revealed by the pattern of resource use was due primarily to foreign trade and credit policies is considerably more tentative than that concerning the existence of the pattern. While this explanation appeared more plausible than any of the others considered, and was consistent with all the information concerning operation of the Chilean credit system and the mechanisms for controlling imports of farm inputs, the analysis was not exhaustive.

Because they are based on analysis of a single sample of farms, none of the conclusions about productivity differences can command the same confidence as that concerning the pattern of resource use. The most important conclusions about productivity are those indicating the amounts of the differences: the conclusion that, except on livestock farms, labor was the most important single source of the differences; and the conclusion that the labor differences were due principally to scale economies in the use of machinery which took hold at farm sizes between 100 and 200 hectares.

Respecting the amounts of the productivity differences, there is little to be said beyond noting that the measurement technique used is a species of a widely recognized genus (the Kendrick method) and that the differences obtained are consistent with the generally held notion that in Chile there is a substantial gap between farms at opposite ends of the productivity spectrum. That for most farm groups the greater part of the productivity differences should be centered in labor may at first glance appear implausible since most of the labor employed on all farms was unskilled. However, labor productivity differences do not simply reflect inherent quality differences in the human material. The quantities and kinds of other inputs with which labor is combined may be much more important, and this in fact seems to have been true in Chile. This particular conclusion rests decisively on the existence of scale economies in the use of farm machinery that became operative between 100 and 200 hectares, and on the absence of rental services of farm machinery. We are satisfied that these services were insignificant in Chile in the late 1950's. The assertion that there are scale economies in farm machinery at 100–200 hectares, however, is less securely based. It rests on the verbally delivered opinions of two CORFO experts, but one would feel much more comfortable were evidence available from in-depth studies.

The most unsatisfactory part of the productivity analysis in our judgment is that part dealing with the role of managerial differences. There is a general presumption among agricultural economists that management accounts for a major part of the productivity differences between farms. Yet the conclusions reached in the last chapter assign a distinctly secondary role to management. This would cause no concern were the conclusion based on a careful evaluation of the management contribution. For want of a measure of management, however, no such evaluation was possible. It is true that such partial indicators of management quality as were available did not suggest marked differences among the farmers sampled. However, these indicators by no means would reveal all the dimensions in which management could manifest itself.

The unsatisfactory analysis of the role of management combined with the less than overwhelming evidence of the existence of scale economies in the use of machinery means that the conclusions regarding the sources of productivity differences are quite tentative. Management may well be more important and machinery economies less so than indicated.

Perhaps the most important general result of the analysis is to alter in some important respects the commonly accepted image of Chilean agriculture. That the performance of the sector could have been considerably better in the 1950's is not at issue. However, the explanation given here for that inferior performance differs rather sharply from any of those generally advanced. Relative prices of inputs and outputs do not appear to have posed a serious obstacle to greater use of modern inputs, even though these prices moved unfavorably over the course of the decade. Apparently as late as 1959 fertilizer still was favorably priced in relation to its rate of return, even though there was no fertilizer subsidy paid in that year. Moreover, in the first part of the decade the effective price of machinery purchased with CORFO loans was substantially less than the nominal price because of the failure to adjust these loans for inflation.

Those who refuse to accept the unfavorable price relations explanation usually turn instead to the land tenure system. We concluded that there was much truth in the assertion that the system has worked to deny adequate educational opportunities to the great mass of farmers and farm workers, and this undoubtedly has adversely affected the pace of technical innovation. In addition the terms of leasing arrangements must have seriously weakened the incentives of tenants to undertake improvements and expand production. Finally, the linkage of land ownership to credit availability effectively denied most farmers the resources needed to acquire modern inputs.

The evidence does not support the assertion, however, that the land tenure system so weakened the innovative spirit among large landowners as to make them disinterested in improving their operations. That this may have been true of some large holders is not denied. But there was considerable technical innovation in the 1950's, much of it financed by credit, of which large landowners got the lion's share. The conclusion is inescapable: as a group large holders were innovators.

Given the continuing debate in Chile concerning land reform, it is natural to ask what relevance the above conclusions have for this debate. We believe they support the notion that land reform will have a favorable impact on the pace of technological progress. In the short run this should work primarily through improved incentives, provided that recipients of land also obtain greater access to modern inputs. In the longer run, more equal distribution of land and other resources ought to increase the political power of the mass of farmers, thereby enabling them to obtain a greater share of the nation's educational resources.

This is not to deny that land reform may have some unfavorable short term impact on the incentives of large landowners. These negative effects, however, are due principally to uncertainties arising from the process of instituting the land reform. Once the reform is well established and its modus operandi understood on all sides, the disincentive effects on large landowners ought to be largely eliminated.

The finding that large farms were more productive because they enjoyed machine economies is only superficially in conflict with the assertion that more equal distribution of land would promote technological progress. The important thing is that farmers should have adequate access to relatively cheap machinery services. To achieve this it is not necessary that each farmer own his own machinery. The necessary services can be provided on a rental basis by either private or public organizations operating in the principal agricultural regions. Properly equipped and managed, such organizations could bring machinery operations to even very small farmers, providing them with the same productivity advantages now available mostly to larger farmers. Hence more equal distribution of land need not require the sacrifice of economies in the use of machinery.

All the preceding analysis and the various conclusions are directly relevant, of course, only to Chile. It is believed that both the analysis and the conclusions would be useful to Chilean agricultural planners, or at least that they would have been at the beginning of the 1960's had they been available then. The procedures followed in developing the analysis are not limited in their relevance to Chile, however. In this respect the study has some things of interest to say to the wider audience of agricultural planners wherever they may be, but especially to those working in the less developed countries. The rest of this chapter deals with these more generally applicable analytical procedures.

CHARACTERISTICS OF AGRICULTURAL PLANNING

It is useful to begin by considering the process of agricultural planning in the abstract. What are its principal characteristics? What is it that agricultural planners are trying to do? What sorts of questions do they ask and how do they go about seeking solutions? Gittinger's survey of the agricultural planning literature provides a useful guide in searching for answers to these questions.[1] The literature now is very substantial — Gittinger cites 423 different sources, although he claims his coverage is not comprehensive. No attempt has been made here to survey all this literature, much of which in any case touches agricultural planning only very tangentially. Moreover, a sampling indicates rather general agreement on the principal characteristics of planning in this field. There is a consensus that the principal broad objective of agricultural development planning is to accelerate the expansion of agricultural production and productivity. Other frequently mentioned secondary but still important objectives are to raise standards of nutrition, achieve greater equality in the distribution of wealth and income among rural households, and increase the participation of farm families in the economic, social, and political life of the country.

The function of an agricultural development plan is to detail and quantify these broad objectives to the extent possible and to indicate the various public and private measures which must be taken if the objectives are to be achieved. A logical first step is a diagnosis of existing conditions in the agricultural sector. Ahumada has asserted that the purpose of the diagnosis is twofold: (1) to describe and analyze the existing situation in the country, region, or sector, and (2) to analyze how this situation came about.[2] This is as good a general statement as one is likely to find. It simply states the common sense notion that in order to know where we can go and how we can get there we must first know where we are and how we arrived here.

With a good diagnosis in hand the planner is prepared to move to the next stage: preparation of demand and production targets by means of projections. Demand projections usually include at least three components: (1) domestic demand for food; (2) domestic demand for nonfood agricultural products; (3) export demand for food and nonfood products. Less commonly, demand to support buffer stocks or inventories also is projected. The product detail in which these projections are made will vary from country to country, but since in most less developed countries agricultural production is dominated by a relatively few commodities, some amount of product detail is feasible and commonly encountered.

---

[1] J. Price Gittinger, *The Literature of Agricultural Planning* (Washington: National Planning Association, 1966).

[2] Jorge Ahumada, *Teoría y programación del desarrollo*, First Series of the *Cuadernos* of the Instituto Latinoamericano de Planificación Económica y Social (Santiago, 1967), p. 113.

Projections of production are usually based on past trends with adjustments to allow for the effects of proposed policy measures. In a properly formulated plan, projected demand will necessarily be equal to the projection of production plus net imports, with policy perhaps serving as the mechanism for achieving this equality. That is, the projections may reflect the play of policy measures both in stimulating production and in restricting demand in such fashion as to achieve *ex ante* the equality which necessarily will be achieved *ex post.*[3]

Thus the agricultural planning process is made up of three basic activities: (1) diagnosis of the present and recent situation in the agricultural sector; (2) projections of the demand for and supply of agricultural commodities and of whatever other targets may be included in the plan, such as amount of land to be brought under irrigation, number of farm jobs to be created, acres of land to be distributed under a land reform program; (3) an indication of the various policy measures which must be undertaken to achieve the projected targets.

RELEVANCE OF THIS STUDY TO THE PLANNER'S TASK

The approach taken in this study provides the planner with a concept of the process of agricultural development and a technique for analyzing it which should be useful in dealing with each of the major aspects of his job: diagnosis, projections, and policy formulation. The basic concept is that the rate of increase of agricultural productivity and output depends upon the pace of invention and adoption of new inputs, and that the speed at which this occurs is set by institutional factors governing the flow of resources into these activities and affecting the technical abilities and incentives of farmers to innovate. The analytical technique employed here is derived directly from this concept of the process of agricultural development. It identifies and assesses the relative importance of various inputs in explaining both the growth of output over time and differences in productivity between farms at any given time. Secondly, it identifies those parts of the institutional structure of greatest importance in explaining the performance of agriculture. Let us examine in more detail the relevance of all this to the agricultural planner.

Diagnosis

A good diagnosis should provide the planner with answers to three basic questions: (1) How has the agricultural sector performed? (2) How has it performed in relation to its potential? (3) What have been the principal obstacles causing actual performance to fall short of potential? Each of these

---

[3] "Policy" as used here means all governmental measures taken to achieve the plan objectives for the agricultural sector. Hence the term includes price, tax, and foreign trade policies as well as extension activities, public investment projects, land reform efforts, and so on.

general questions consists of a number of more specific ones which require treatment in some detail.

Actual performance of agriculture. There are many dimensions to performance, but those of most relevance to the planner are two. The first is the growth of production over time, both total and, if possible, broken down by principal products and geographical regions. Since the production growth rates of various commodities may be quite different, the planner must necessarily examine performance on a commodity basis if he wishes to formulate policies appropriate for particular commodities. However, if the necessary commodity detail is not available, the planner may still be able to draw some useful conclusions from analysis of the growth of total production or of broad commodity aggregates. In the present study, for example, it was possible to make some headway in analyzing the performance of Chilean agriculture in the 1950's even though the commodity classes were as broad as "livestock," "annual crops," and "nonfruit" farms.

The analysis of performance on a regional basis will be desirable whenever producing regions differ markedly with respect to climate and soils, which will be the case in most countries.[4] Once again, however, it is noteworthy that the analysis of the growth of production in Chile in the 1950's proved fruitful even though only slight account was taken of regional differences in performance. This does not mean, of course, that the conclusions drawn from that analysis apply with equal force to all producing regions, clearly an unjustified inference. The result does show, however, that *some* aspects of performance may well be sufficiently common to all regions to create a nationwide pattern, and that analysis focused at the national level can discern and draw useful conclusions about the contours of this pattern.

The second major dimension of performance of concern to the planner is the sources of output growth. This follows directly from the concept of the development process underlying this study. If the growth of production and productivity depends upon the rate of invention and adoption of new inputs, the planner's attention is inevitably directed to these inputs when he undertakes to evaluate past performance. He will wish to identify the inputs which have contributed to output growth, to measure the increase in the amount employed of each, and to estimate their relative contributions to the growth of production. He will then be able to determine the marginal rate of return to each input, information of considerable value in explaining both the growth and pattern of input use as well as the growth of output.

In the diagnosis of the performance of Chilean agriculture in the 1950's two methods were used to identify and measure the contribution of the various sources of output growth. One made use of uncertain and fragmentary information concerning the increase in employment of various in-

---

[4]Ahumada, *Teoría y programación*, p. 115, emphasizes the importance of regional differences in agricultural planning.

puts over the course of the decade. Despite the serious limitations in the data, this technique produced a number of important conclusions about the sources of output growth and their marginal rates of return. This suggests that the technique may provide the planner with a useful diagnostic tool, even under conditions of severe data deprivation.[5]

The second technique relied upon was analysis of production functions to identify the various inputs and measure their contribution to output. The functions were based on sample data for a single crop year, and hence revealed nothing about the sources of output growth over time. Rather they provided a snapshot of performance at the end of the 1950's. Nonetheless, this snapshot, taken together with the results of the analysis of performance over the decade, provided information of immediate relevance to the planner. It permitted a check on the reasonableness of the estimates derived for the nation as a whole from the time series analysis. But more important, the production function analysis yielded more reliable estimates of rates of return to more inputs than would be derived from the study of time series.[6] These estimates would serve the planner in a variety of ways, not least in helping him to deal with the second major component of the sectoral diagnosis: determination of potential output.

Potential compared with actual output. In all the language of economics, there probably is no more slippery concept than that of "potential" output. Potential implies a limit, and a limit implies constraints on the expansion of production. The elusiveness of the concept of potential output stems from the fact that different sets of constraints can be stipulated, each of them setting a different level of potential output. One might define the potential output of a plant, for example, as the maximum amount the plant can produce annually, given an unlimited supply of free variable inputs. This is the output achieved when the marginal products of the variable inputs are zero. The only constraint on production in this case is the stock of capital in the plant. At any given time this stock is fixed. However, with time it can be increased or diminished, with a corresponding increase or decline in potential output. Given sufficient time, therefore, the capital stock of the plant no longer acts as a constraint on output. Instead the limiting factors are those

---

[5]The technique does impose certain minimum data requirements, however. Reasonably reliable estimates of production are essential, as are estimates of the increase in the quantities of various inputs employed. There doubtless are many countries less able to meet these requirements than Chile. Moreover, in the analysis of Chilean performance we were assisted by the fact that the quantities employed of two major inputs – land and labor – did not change much over the period studied. This narrowed considerably the range of speculation concerning the contribution of other inputs, giving this speculation more point.

[6]As noted above, the time series analysis indicated that the quantities of land and labor employed changed little, if any, over the decade of the fifties. Hence, marginal rates of return to these inputs could not be calculated with this technique. Moreover, the time series data, particularly those for the capital stock, were judged to be inferior to those derived from the sample survey.

which determine whether the capital stock of the plant will be increased or diminished, and if so, by how much. Hence, if one analyzes the potential of the plant over, say, a decade, attention will be focused on those factors determining the flow of "fixed" capital into or out of the plant. However, if potential in a single year is of interest, then the stock of capital existing in that year is the crucial factor.

If the concept of potential output is to be of any use to the planner, it must have normative content, that is, the concept should carry the notion not only that given certain constraints a certain level of output *could* be achieved; it must also imply that this level of output is socially desirable. Recognition of the normative element in the concept of potential output implies acceptance of a behavior rule – that resources shall be disposed in such fashion as to maximize the social value of the product yielded by them – and indicates that potential output cannot be specified independently of the prices of inputs and outputs (assuming these prices reflect marginal social costs and values).

These points can be made clear by resorting to the previous example. For periods of a year or less, potential output was defined as the maximum amount the plant could produce annually, given an unlimited supply of free variable inputs. At this level of output the marginal products of the variable inputs are zero. Given the assumptions, this is the most socially desirable level of output. Since the variable inputs are free, it pays to use more of them as long as they add anything to output. Even though they are free, however, it obviously does not pay, given the behavior rule, to use so much of them that their marginal product becomes negative, for then output is less than it otherwise would be.

Thus, given the assumptions, the definition of potential specifies a socially desirable level of output. The assumptions, however, fly in the face of a basic fact of life: most variable inputs are not free. In this case the definition of potential must be changed. For any period of a year or less it now reads: that level of annual output achieved with the given stock of capital when the marginal social products of inputs are equal to their prices (assumed to equal their marginal social costs). This defines the most socially desirable level of output for the plant. At any lesser output the marginal social products of the inputs would be greater than their prices while at a higher level of output the reverse would be true. In either case the total social product is less than it "ought" to be.

Obviously the plant's potential output is less when prices of inputs are taken into account than when they are assumed to be free goods. Moreover, there now is a different level of potential output corresponding to every possible combination of variable input and output prices. To complicate matters even more, if one wishes to talk about potential over a period long enough to permit changes in the "fixed" capital of the plant, then the prices and marginal social products of capital goods also must be taken into account.

In assessing the performance of the agricultural sector the planner must be aware of these complexities in the concept of potential output. In particular, he should recognize that potential cannot be defined independently of the prices of variable inputs and outputs, and that generally the longer the period of time under consideration the larger is potential output because with more time fewer inputs are fixed.

The analytical techniques employed in this study permit the planner to make some headway in estimating potential agricultural output. As a first approximation he can define potential as that level of output achieved when, for every input, prices and marginal products are equal. By means of production functions he may then derive estimates of the marginal products and compare these with input prices. This process will indicate whether farmers could profitably have used more inputs, whether they employed too many, or whether they were using just the right amounts. In the first case, actual output would have fallen short of potential; in the second, too much would have been produced (i.e., some of the inputs could have been more profitably employed elsewhere in the economy); in the third, potential and actual output would be equal.

This concept of potential output is consistent with that discussed above: it takes account of prices of inputs and outputs and it stipulates the behavior rule which assures that inputs will be employed in such fashion as to maximize their social product. Implicitly this particular concept is long run, i.e., it defines a level of output which could be achieved if all inputs were variable, including "fixed" capital.

This is the technique that was used in the analysis in Chapter 4 of the performance of the 177 nonfruit farms in Chile's O'Higgins Province. The analysis indicated that those farms, taken as a group, could have profitably employed considerably more of all nonland inputs, except livestock. In the context of the present discussion, they were operating well under their output potential. This kind of information clearly would be of considerable planning interest. However, it has the limitation that the indicated level of potential output is in effect an average, i.e., it shows how much those farms *as a group* could have profitably produced, given existing input and output price relations. While this concept of potential is useful, the planner may want another more demanding standard of comparison, namely, a measure of potential defined as that level of output which would have been obtained if all farms had been as productive as the top X per cent.

A modification of the technique employed in Chapter 5 for the analysis of more and less productive farms would permit the planner to derive an estimate of potential output corresponding to this second definition. The first step would be to classify farms according to their total productivity ratios, in the manner described in Chapter 5. Next, a production function would be computed for the top X per cent of farms. Third, the inputs actually employed by the whole group of farms would be substituted in the production function for the top X per cent. The result would be an estimate of the production that would have been achieved by the whole group of farms, given the inputs actually used, if the average productivity of the group had been as high as that for the top X per cent. Finally, a comparison of the prices and marginal products of the various inputs would indicate the changes in the quantities of inputs required to achieve potential output, given the production function of the top X per cent of farms.

Earlier in this discussion it was stated that the suggested techniques permit the planner "to make some headway" toward estimating potential production and that the results obtained would be treated as a "first approximation." This cautious language was not accidental. It reflects some important reservations concerning the estimates of potential output yielded by the suggested techniques. These reservations apply with equal force to both definitions of potential discussed above.

In the first place, both definitions assume that input and output prices faithfully represent the social values of resources and products. This may not in fact be so, for a variety of reasons long recognized by economists. Some activities may involve social costs and benefits which, being social, are ignored in the investment and production decisions of private producers. Chemical wastes discharged into streams may pollute the water to such an extent as to make it unusable for downstream irrigation, thus reducing potential agricultural production. This is a genuine social cost of water pollution, but it is not a cost relevant to the operations of the plants responsible; consequently, it will not be reflected in the prices of their products. These prices, therefore, will be lower and agricultural prices will be higher than they would be if the social costs of pollution were properly taken into account.

As another example, the social return to additional investments in development of improved seed varieties may be very high, *if* complementary investments are made in farmer education, extension activities, expanded agricultural credit facilities, and improved marketing institutions. These complementary investments, however, will lie outside the purview of private producers of improved seeds. Consequently, the return to increased investments in seed development as perceived by them will be less than the social rate of return, resulting in a less than socially optimum amount of investment in this activity.

Other cases of "market failure" arise from the existence of monopolistic elements in output and input markets. In these circumstances prices will not properly measure the marginal social costs of resources or the marginal social value of outputs. Given the rigid institutional structures in many less developed countries and the quite unequal distribution of political, social, and economic power, considerable and persistent departures from competitive conditions are to be expected in many markets.

The significance of all this for the agricultural planner is that the outputs achieved when market prices and marginal products of inputs are equal do not represent the socially desirable amounts. In this case, the estimates of "potential" production lack normative significance and hence provide no guidance to the planner in diagnosing the performance of the agricultural sector.

If the planner has some idea of the extent of the divergence between actual prices of products and inputs and their real social values, he can make at least a rough adjustment, substituting estimates of these real values in calculating marginal products and costs of inputs. For example, he may judge that because of price controls or consumers' imperfect knowledge of relative nutritive values, the price pattern for foods does not adequately represent the relative social values of different foods. He may, therefore, stipulate a price pattern which more closely approximates social values and recalculate the marginal products of the inputs used in the production of the affected commodities. Comparison of this set of marginal products with input prices (assuming the latter represent real social costs) would indicate the changes in quantities of inputs required to achieve genuine potential production.

Of course, prices of inputs may also diverge from real social costs, in which case the planner's task in estimating potential output is even more difficult. Nevertheless, the effort still would be worthwhile, even though the final product was known to be far from perfect. At the very least, the exercise will show whether, with the given set of input and output prices, resources were allocated so as to maximize production. This can provide to the planner some useful insights into farmer behavior and perhaps indicate something about the operation of product and input markets.

A second reservation about the suggested techniques stems from the fact that, in principle, they require data from all important producing areas of the country, and the data must include all outputs and all inputs which contribute significantly to production. The importance of recognizing regional differences has already been noted, and it is obvious in any case. One of the limitations of the analyses in Chapters 4 and 5 is that they are based on data from only one region of Chile, a part of the Central Valley. Consequently, the conclusions drawn, insofar as they apply to the country as a whole, must be more tentatively held than would be the case were they based on data taken from a large number of regions. Of course, where resources for planning research are adequate, data can be collected from all important regions. In this case, the planner will be in a position to analyze performance region by region, clearly a desirable thing to do where regional differences are marked.

The usefulness to the planner of the suggested techniques also depends upon the adequacy of the data in covering all significant outputs and inputs.

In areas where large amounts of agricultural production escape the market and few inputs are purchased, the production function approach may prove unworkable. Of course, if one could measure the quantities of outputs and inputs and had some reasonable basis for imputing unit values to each, the analysis could go ahead. As a practical matter, however, obtaining this information is likely to cost more than it is worth.

Even where most production enters the market and most inputs are purchased there may still be problems, particularly on the input side. The difficulties in measuring management inputs have been touched upon several times in the course of this study, and they bear reemphasis here. The point is important because the quality of management can be increased by allocating more resources to farmer education and extension services. However, the optimum increase in resources devoted to these uses cannot be accurately determined if the planner does not know the returns to improved management quality. His inability to include management in production functions makes it quite difficult to estimate these returns.

It also is highly desirable for the planner to have estimates of the rates of return to infrastructure services, yet these inputs can seldom if ever be included in farm-level production functions.[7] The contributions of roads, extension services, and irrigation systems obviously are of major importance. Only the last is likely to be measurable, however, and even it may not be. In Chile, for example, where 50 per cent of agricultural production originates on irrigated land, inputs of irrigation services – water – are not presently measurable for want of data.

These several reservations concerning the use of the suggested techniques for measuring potential output are just that: reservations. They are emphasized to avoid conveying the impression that the techniques promise to solve all the planner's problems of diagnosis. Obviously, they do not. Still, used with due awareness of their limitations, they offer a relatively economical means of presenting the planner with highly relevant information concerning actual and potential performance of the agricultural sector. While nothing more is claimed for them, that seems quite enough.[8]

Obstacles to achievement of potential output. In the discussion above of the two definitions of potential production – one based on the production

---

[7]This writer knows of no instances in which they have been.

[8]The assertion that these techniques are "relatively economical" is based on the fact that they can make use of data collected through sample surveys or perhaps from agricultural censuses, the latter of which would be undertaken in any case. I have no information on the cost of the sample survey used in this study, but it probably did not run over $10,000. To be sure, samples from other regions of the country would be necessary to reach firmly grounded conclusions applicable nationwide. Even so, the total cost of such surveys, in relation to the value of the information they would yield, would probably appear quite reasonable.

function for the whole group of farms and the other on the function for the top X per cent – no attention was given to what perhaps seems an obvious question: for the planner, which definition is the more meaningful for diagnosing past performance? The answer is, it depends. It depends upon the planner's judgment of the relative importance of the various obstacles to the achievement of potential and also upon the term of the plan under consideration. In general, the longer the term of the plan, the greater the possibility of removing any given set of obstacles. The longer the term of the plan, therefore, the more likely that the planner will find the second definition of potential more useful than the first. The reason is that it is likely to be more difficult to raise the average productivity of all farms to the present level of the top X per cent than it is to reach the socially desirable level of output achievable with the existing average production function for all farms. The problem is not necessarily that the existing average production function embodies a wholly primitive technology while that of the top X per cent of farms is technically advanced. This may be the source of the difficulty in some countries where agriculture is characterized by marked dualism, a primitive subsistence sector existing side by side with a modern export sector. But in other countries – Chile, for example – the situation appears different. The average production function is by no means technically primitive, as that term is commonly used in development economics. Fertilizers, pesticides, and farm machinery, particularly the first two, are relatively widely used. They are by no means reserved mainly to the top X per cent of farms. Nonetheless, raising the average productivity of all farms to the level of the top X per cent is likely to be more difficult than achieving socially desirable output with the existing average production function. The reason is that the high productivity of the top X per cent of farms typically is due to conditions which are relatively harder to change than are those inhibiting realization of the output potential of the existing average production function. Above all, the top X per cent of farmers generally will have considerably more general and technical education than is reflected in the average production function. To raise the average level of educational inputs to that presently enjoyed by the top X per cent is inevitably a time consuming process. In addition, the top X per cent of farms generally will be substantially larger than average. To the extent that the Chilean experience is any guide, this circumstance conveys significant advantages in acquiring credit and also in the use of machinery. Given the rigidity of land tenure systems in most parts of the world, the size advantage is not something the planner can expect to alter quickly. Even if the disadvantage of small farmers in the use of machinery can be alleviated by the expansion of machinery rental services, this is likely to be a time consuming process. The economical provision of such services on a large scale requires a highly organized operation involving not only the equipment itself but substantial numbers of trained operators and mechanics, repair and maintenance facilities, accountants, and other office personnel, and a high level of managerial skill. In the typical underdeveloped country all of these are high-cost resources, the supply of which cannot be readily increased.

Thus the obstacles to raising the average productivity of all farms to the level of the top X per cent are likely to be deep-rooted and only slowly overcome. By contrast, the obstacles to realization of the output potential of the existing average production function probably will appear less severe. The problem is essentially one of allocating more or fewer (but almost surely more) resources to agriculture. In countries where input markets are poorly organized, agricultural credit and extension systems primitive, balance of payments problems severe, and government budgets able to do little more than meet current operating expenses, it may prove quite difficult to do this. However, not all less developed countries by any means are so hard pressed in these respects. In Chile, for example, the principal obstacles to an increased flow of resources to agriculture in the 1950's appear to have been a complex of credit, foreign-trade, and price policies. Alteration of these policies to stimulate agriculture clearly lay within the capabilities of the Chilean government. It surely would have been much easier to accomplish this in the decade of the 1950's than to significantly increase the amount of general and technical education available to farm people or to achieve a far-reaching land reform.

The preceding discussion indicates that the most meaningful definition of potential output will depend upon the term of the plan to be formulated and upon the planner's perception of the nature of the obstacles impeding the expansion of production. The concept of development underlying this study states that the locus of these obstacles is in the institutional structure, particularly that part governing the invention and adoption of more productive inputs. In completing the diagnosis of performance, therefore, the planner must turn to analysis of these institutions.

Institutional Obstacles to the Realization of Potential Output

To guide this examination, the planner may find it useful to divide the problem into three parts: (1) those institutional forces governing the *technical availability* of fertilizers, pesticides, machinery, water, and improved seed varieties adapted for use under the range of soil, topographical, and climatic conditions in the country; (2) those forces determining the *technical ability* of farmers to make effective use of these inputs; (3) those forces affecting what we will call the farmer's *economic ability* to effectively employ greater quantities of these inputs. That is, a wide range of inputs technically well adapted to the farmer's specific climatic and soil conditions may have been invented and tested, and he may possess all the technical skill necessary to employ more of them yet be unable to do so for a variety of economic reasons. These would include unfavorable input-output price relations, inefficient organization causing bottlenecks in product and input markets, physical rationing of inputs by public authorities, and insufficient credit for the purchase of inputs.

The first set of institutions are principally those devoted to research on, and development of, inputs, except that for water the public and private institutions concerned with the construction and operation of irrigation systems will be the focus of interest. For the other inputs, probably a host of both public and private institutions will be involved, and the planner will want to make a thorough canvass of the whole range of these activities. Because the market mechanism does not ordinarily permit capture of the full return to basic research, public or nonprofit private institutions are likely to be dominant in this field.[9] The number of such institutions in the typical less developed country is likely to be small, and their activities are likely to be highly visible to the planner. Hence, he is not likely to overlook them.

There may be a tendency, however, to give less attention than is deserved to the efforts of commercial firms in research on and development of new inputs. These may in fact be a significant part of the total effort in this field. The joint ECLA/FAO study of the use of pesticides in Chile noted that commercial companies were responsible for almost all of this type of work on pesticides.[10] Private producers of fertilizers and farm machinery, particularly when these are large firms producing for an international market, also are known to engage in research and development activities with respect to these inputs.

Institutions affecting the technical ability of farmers to make effective use of new inputs are those devoted to education and extension activities. In a sense these institutions are themselves engaged in the production of new inputs, namely human skills. Nonetheless, the nature of their activities is sufficiently different from that of institutions producing fertilizers, machinery, and so on to justify setting them apart for purposes of diagnosis. The importance of the relationship between farmer education and agricultural productivity is now widely recognized by students of economic development. Earlier in this study considerable emphasis was given to an assessment of both the quantity and quality of education available to Chilean farm youth, and it was concluded that the very serious deficiencies in this field must be an important element in the low productivity in much of Chilean agriculture. In a contrasting situation, an FAO study of Japan concluded that the rapid expansion of both general and technical education since the late nineteenth century was a major factor in that country's successful transformation of its agricultural sector.[11] In the report of the MIT Conference on Productivity

[9]Cf. the discussion in Chapter 1, above; also, T. W. Schultz, *Transforming Traditional Agriculture* (New Haven and London: Yale University Press, 1964), Ch. 10.

[10]*El uso de pesticidas.* (See Ch. 2, n. 64.)

[11]"Agricultural Development in Modern Japan" (final section of a paper prepared by a group of experts for the Food and Agriculture Organization for presentation at the World Food Congress, held in Washington, D.C., June 4–18, 1963), p. 18.

and Innovation in Agriculture in the Underdeveloped Countries it is asserted flatly that "the importance of education of farmers as a means of increasing agricultural productivity in both the short- and long-term scale cannot be overemphasized."[12]

The alert planner is not likely to be insensitive to the importance of education. However, lack of data as well as the difficulty of discerning clearly the relationship between different amounts and kinds of education and the acquisition of farm skills may make it impossible for him to go very far in quantifying the impact of educational institutions on agriculture's performance. He is likely to find the going even harder when he attempts to explain why these institutions behaved as they did. For the resources devoted to education and the relative emphasis given to various types of education reflect deeply rooted and diffused cultural values which, precisely because they are so deeply and widely held, may be so taken for granted as to obscure their origins and made difficult the tracing of their consequences. In a sense, education is the medium by which cultures perpetuate themselves; accordingly the educational process reflects the culture's image of itself. A full account of the processes determining the amounts and kinds of education made available to farm people, therefore, would involve the planner in analyses going to the very heart of the value system which holds his culture together, giving it such direction and momentum as it may have.

Such a task obviously lies outside the competence of the typical planner. He may nonetheless be able to make some progress toward a more modest objective. In most countries the allocation of resources to education occurs through political processes, the analysis of which should be within the planner's capabilities, particularly if he can draw on the work of knowledgeable professionals in political science. The distribution of political power and the ways in which it is exerted are after all not wholly occult matters. In Chile, for example, it is apparent that political power is related to land ownership, with the result that the distribution of power in the countryside is highly unequal. In the earlier discussion it was hypothesized that this may account in large part for the relatively small amount and poor quality of education made available to Chilean farm youth in the 1950's.

This clearly is far from a complete explanation of the processes determining the flow of resources to rural education in Chile. Yet it is suggestive, and it illustrates the point that while a full explanation may forever elude the planner, he may yet be able to devise partial hypotheses of considerable diagnostic value.

The third element in the planner's diagnosis of the obstacles to achievement of full potential output concerns those institutions which bear on the

[12] David Hapgood (ed.) and Max F. Millikan (conference chairman), *Policies for Promoting Agricultural Development* (Cambridge, Mass.: Massachusetts Institute of Technology, Center for International Studies, Jan. 1965), p. 109.

farmer's economic ability to use greater quantities of modern inputs. To a large extent these are economic institutions per se: product and factor markets and credit institutions. In the analysis of them the planner – whose training is more likely to be in economics than anything else – will find himself on more familiar ground than in the investigation of research, development, educational, and extension institutions. To the extent that the analysis of actual and potential performance follows the procedures employed in this study the planner will already have a notion of which set of economic institutions is most relevant in explaining performance. For example, the production function analysis will indicate whether product and input price relations favored employment of greater quantities of modern inputs than were in fact employed. If it appears that price relations were favorable, the planner can put this factor to one side and concentrate on nonprice forces bearing on the farmer's production decisions. One possibility is that found to be operative in Chile in the 1950's, namely, government foreign-trade and credit policies. The former are likely to be important in any country which depends substantially upon imports for its supply of modern inputs, while credit policies – indeed, the performance of the entire credit system – almost inevitably will be of major significance everywhere. There are few farmers in a position to finance large-scale purchases of modern inputs out of their own resources. Most are critically dependent upon the availability of credit. Hence, study of the performance of credit institutions should be an essential part of the planner's task, whatever else it may include.

The supply of inputs may also be inhibited by faulty organization of input markets. How do these operate? Who controls them? Why are they not responsive to the excess of demand over supply implicit in input-output price relationships? Are these markets sensitive to the seasonality of the demand for agricultural inputs, and if not, why not?

It may be that the problem lies in the inefficient functioning of product markets. Are there such bottlenecks in the process of moving commodities from the farm gate to the final consumer that distribution costs would mount steeply were the quantity produced much greater? These bottlenecks may be owed to weak transportation systems, inadequate or poorly located storage facilities, escalating losses due to spoilage when the amount brought to market rises much above present levels, fragmentation of retail markets resulting in small-scale, high-cost operations, and so on.

The institutions governing the ownership and control of land may, of course, have a major impact on the incentives and economic ability of farmers to employ greater quantities of modern inputs, even when relative input-output prices are favorable. To the extent that the land-credit-input nexus prevails, only those farms with secure title to an adequate amount of land may innovate. Where leasing agreements are predominantly short term and assign the landlord a major share of the fruits of innovation, landless farmers will lack incentives to employ more modern inputs, even in the event that they could find the wherewithal to purchase them. Finally, there is the possibility that the land tenure system weakens the incentive of landlords, particularly large ones, to behave like "economic men." This, it will be remembered, was the principal hypothesis of the CIDA Report in explaining the presumed lack of technological dynamism in Chilean agriculture. While the hypothesis was found to be unconvincing in the particular case of Chile, it nonetheless is one that, in general, planners should consider, at least until sufficient evidence has been compiled to demonstrate its general invalidity. In any event, the impact of the land tenure system on innovation by way of the credit system and leasing arrangements is likely to be of sufficient importance to command the planner's attention, even if he decides the supposed effect on incentives of large landowners is a myth.

The analysis of Chilean agriculture in the 1950's indicated that input-output price relations were *not* the principal obstacle to higher production. However, this may be the case in other countries, or in Chile at some other time. How would the planner detect this? The procedures followed in the study of the Chilean situation would yield much if not all the needed information. Analysis of production functions would provide estimates of the marginal products of the various inputs. If comparison of these with input prices showed they were equal, the conclusion would follow that price relations were impeding the expansion of production.

Here the planner would confront an apparent paradox. If input prices and marginal products are equal, then resources are employed to maximum efficiency. In this case, in what sense are price relations impeding the expansion of production, since to produce any more would require inefficient use of resources? Yet it is evident to the planner and to the community at large that the present level of output is "inadequate" in some sense and "ought" to be increased.

The paradox stems, of course, from weaknesses in the received concept of economic efficiency. As was noted earlier in this chapter, the concept requires the assumption that input and output prices measure the real social values of resources. The planner implicitly rejects this assumption when he concludes that the existing level of output is "inadequate" even though input prices and marginal products are equal. For if input and output prices measure real social values, then no more inputs *should* be employed nor output produced, in which case the existing level of production must be adequate. Performance which is as it "should" be cannot be inadequate.

The presumed situation is by no means far-fetched. Indeed, Schultz has argued that it tends to exist wherever primitive agricultures are found. Such agricultures are marked by little if any technical innovation, either because input-output price relationships for modern inputs are unfavorable, or because farmers lack the technical know-how to manage these inputs effectively, or both. Hence, farmers work only with traditional inputs and have had ample time to adjust their use of them so as to equalize their prices and rates of return. Within the economic and technical limits of their situation, they are "efficient" producers.

Nonetheless, Schultz does not conclude, nor will the planner, that their performance is "adequate" in the sense that no effort should be made to improve it. The thrust of this effort will be either to raise output prices or to lower the cost of inputs, or perhaps both simultaneously. Unless the planner knows that output prices have been suppressed by administrative controls his most practical course may be to assume that these prices roughly approximate true social values and to devote himself, therefore, to the study of why the costs of modern inputs are so high. This inevitably will lead him to study the institutions governing the flow of resources not only into the invention and production of new inputs but also into educational and extension activities.

In countries where output prices are subject to official controls, the planner will be obliged to examine the mechanisms by which these controls have been enforced and to estimate the extent to which they have prevented prices from rising. Unless there are private monopolistic elements in the production and distribution of these outputs the planner may be justified in concluding that as a result of the controls, prices understate the true values of the various outputs. He may believe, therefore, that it would be legitimate to alter or remove price controls in order to induce farmers to expand production.

Summary of Diagnosis

The procedures sketched above would provide the planner with a useful diagnosis of the performance of the agricultural sector. Through a combination of times series analysis and cross-sectional study of production functions he would not only have traced the path and measured the amount of output expansion but would also have made considerable progress in identifying the major sources of output growth and in measuring their respective contributions. The analysis of production functions, both for farms in general and for groups of the more productive and less productive, would permit him to make different estimates of potential production against which to measure actual performance. Finally, study of the institutional obstacles limiting the technical availability of modern inputs, and both the technical and economic ability of farmers to use greater quantities of these inputs, would provide the planner with considerable insight into why actual production fell short of potential.

With these several elements of the diagnosis of performance in hand, the planner is ready to move to his next major task: projections of the demand and supply of agricultural commodities and specification of the policies required to achieve them.

DEMAND AND SUPPLY PROJECTIONS

It will perhaps be obvious by now that the procedures followed in this study have nothing useful to say to the agricultural planner about demand projections. The focus of the study has been on the factors determining the availability of modern inputs to Chilean farmers and their technical and economic ability to use them effectively. Demand conditions in principle are among these factors because they influence farmers' incentives, that is, they have a bearing on the "economic ability" of farmers to employ more of various inputs. In the particular case of Chile, however, it appeared obvious that demand conditions were of slight importance in this respect compared with government credit and foreign-trade policies and with the host of conditions affecting the technical ability of farmers. No attention was given, therefore, to the role of demand in the performance of Chilean agriculture.

That this study has nothing relevant to say about demand projections does not mean, of course, that they are unimportant. They are in fact essential to the elaboration of a sensible plan, since they provide critical information about the size of the prospective market for agricultural commodities. This information is important to the planner in projecting potential output because it will tell him something about the likely behavior of commodity prices. As was emphasized in the earlier discussion, potential production cannot be specified independently of the prices of both outputs and inputs.[13]

It will be assumed here that the planner has derived at least a tentative set of demand projections and that he now confronts the problem of projecting production. What guidance will the procedures followed here provide him? The problem is to project potential production. The diagnosis of past performance will have yielded estimates of potential in the previous plan period and will have identified and explained the operation of the principal institutional obstacles blocking achievement of that potential. This information provides the basis for the required projections. Two steps are necessary: (1) the planner must estimate the extent to which the institutional obstacles can be overcome in the plan period; (2) he must estimate the impact of this on production.

The length of the plan period is obviously of crucial importance. Because the removal or weakening of institutional obstacles takes time, projections of potential output will in general be less the shorter the period of the plan. However, given this period, the planner must judge the likelihood that the various institutional obstacles can be overcome. He almost surely will conclude that those obstacles limiting the flow of resources into education of

---

[13] There is general agreement in the planning literature that techniques for projections of demand for agricultural commodities are considerably more advanced than those for supply projections. For a sophisticated example of the application of these techniques plus a commendably clear and concise discussion of them, see Food and Agriculture Organization, *Agricultural Commodity Projections to 1970* (Rome, 1962).

farm people will be more difficult to weaken or remove than those restricting the economic ability of farmers to use greater quantities of modern inputs. As already noted, the former set of obstacles is rooted deeply in the culture of the society. The economic ability of farmers to use modern inputs can be significantly increased, however, by such relatively simple measures as, for example, expanded credit facilities, increased production or imports of fertilizer, input subsidies, or removal of price controls on agricultural commodities. The relative strength of the obstacles to greater technical availability of modern inputs cannot be so readily judged a priori. To be sure there already exists a wide range of such inputs, but most of them were developed in the industrialized countries. Work on the adaptation of these inputs to conditions in specific less developed countries is far less advanced. Countries lacking capacity to undertake such work conceivably could find this a major obstacle to technical innovation in agriculture, an obstacle comparable in strength to those limiting the technical ability of farmers to use modern inputs effectively. In general, however, this limitation is more likely to apply to the technical availability of improved seed varieties than to fertilizers, pesticides, and machinery. Development of reliable new seed strains requires extensive testing in both laboratory and field, an activity which typically will require some years. Already available fertilizers, pesticides, and machinery, on the other hand, appear to be much more readily adaptable for use under a wide variety of climatic, soil, and topographic conditions.

Hence with relatively short planning periods, say five years or less, the planner probably will conclude that the institutional obstacles limiting the technical availability of inputs and the economic ability of farmers to use them effectively are considerably less than the obstacles limiting the farmers' technical ability in this respect. Consequently, he will find the estimate of potential based on the existing average production function more relevant than that based on the production function of the top X per cent of farms. Given the relatively short term of the plan, the latter simply is not attainable whereas the former is, or is so in large part, provided the planner has correctly judged the strength of the obstacles limiting the technical availability of inputs and the farmers' economic ability to use them.

The existing average production function will not only provide the planner an estimate of potential production for the next plan period it will also indicate how this potential can be achieved, i.e., it will tell him which inputs must be increased and in each case by how much. The analysis of the production function for the 177 nonfruit farms in Chile illustrates these points. By means of that function we were able to estimate the amounts of various inputs which could have been profitably employed in addition to those actually employed and to derive the substantially greater output corresponding to the increased quantities of inputs. This characteristic of production functions — that given the existing set of input and output prices, they yield not only an estimate of potential production but also the specific quantities of each input required to achieve that potential — makes them particularly valuable tools to the planner. For not only must he project potential, he must also recommend policies for the achievement of it. Knowledge of the required input quantities is invaluable in shaping these policies. If the planner knows that he needs X per cent more fertilizers, Y per cent more machinery, and so on for each other input, and if from his diagnosis he can judge the strength of the institutional obstacles limiting acquisition of these greater input quantities, his ability to suggest effective policies is far greater than if he lacks this information.

In the literature on agricultural planning the possible uses of production functions for projecting output have not gone unnoticed. In his authoritative survey of the existing state of the art, Ojala focuses particularly on this use of such functions.[14] He observes that they are not likely to be very reliable instruments because past studies have shown that increases in the "conventional" inputs of land, labor, and capital account for only a relatively small proportion of the total increase in output, the rest being wrapped up in a residual attributable to time or "technological change." He concludes that since production functions cannot adequately explain past changes in output they "can hardly be used to project future changes."[15]

Ojala's critique carries weight to the extent that production functions fail to include all relevant inputs. However, as the procedures followed in this study show, the inputs need not be limited to the broad aggregates — land, labor, and capital. As was emphasized in Chapter 1, "technological change" is nothing more than the substitution over time of more modern inputs for older ones, where human skills are included among inputs. To the extent that this process of substitution can be reflected in the production function — and differentiation of inputs according to quality will do this — the function can be used to explain changes in output over time, i.e., the residual attributable to "technological" change can be made very small. An empirical study by Griliches using U.S. data has shown how this can be done.[16] To be sure, not many countries can at present provide the quantity of data available in the United States, and Griliches is a very scarce resource. Nonetheless, as the present work on Chile demonstrates, the data on agricultural inputs are not necessarily as limited as Ojala implies, nor need every country possess its own Griliches so long as it has access to Griliches' work.

This is not to say that production functions have no limitations as techniques for projecting potential production. There are in fact some very real

[14] E. M. Ojala, "The Programming of Agricultural Development," in H. M. Southworth and B. F. Johnston (eds.), *Agricultural Development and Economic Growth* (Ithaca: Cornell University Press, 1967), p. 570.

[15] Ibid.

[16] Zvi Griliches, "Research Expenditures, Education and the Aggregate Agricultural Production Function," *The American Economic Review*, LIV, Dec. 1964.

ones. The most important stem from the fact that the functions calculated by the planner reflect (a) a particular pattern of input and output prices; (b) a particular pattern of production (this pattern being related to the price pattern in (a)); and (c) a particular level of "average" technology. The projections of potential output obtained from these functions implicitly assume that these three sets of conditions will not change significantly over the plan period. A moment's reflection will indicate why this is so. If, for example, the price of fertilizers were to rise sharply over the plan period in relation to prices of agricultural commodities, the most economical amount of fertilizer would be substantially less than if the price change had not occurred. If the planner failed to allow for this change in projecting potential output, his projection would overstate both potential output and the needed amount of fertilizer.

The assumption that the production pattern does not change significantly is necessary because the existing average production function is just that – an average. It represents a set of functions peculiar to particular commodities. There may be significant differences between these commodity functions; hence, if the relative importance of the several commodities changes very much, the existing average production function will no longer adequately represent the new pattern of production.

The requirement that there be no significant change in technology is almost too obvious to require comment. The existing average production function measures the net impact on production of combining certain kinds of inputs. Technological change involves the substitution of a different set of inputs for the old ones, where "different" is defined to mean different with respect to the impact on production. By definition, therefore, technological change involves the shift from one production function to another.

The use of the existing average production function for projecting potential output rests, therefore, on three basic assumptions, any of which may prove wrong. The pattern of input and output prices and of production probably will change in response to ordinary market forces as well as deliberate government policies. Suppose, for example, that projected total demand is substantially greater than the existing level and that the pattern of demand is expected to change in response to rising per capita income. The effect will be to sharply increase the pressure on resources generally, especially those required for production of the products for which demand is increasing most rapidly. Unless there is substantial general unemployment of resources, which is unlikely, or unless the supply curves of all resources can be shifted rapidly to the right – which is also unlikely in view of existing institutional obstacles – then the prices of some and perhaps many inputs will rise, although not all in the same proportion. At the same time, it may be decided as a matter of public policy to lower the prices of some inputs – perhaps through a subsidy – in order to stimulate greater use of them. These changes in input prices in turn will affect the relative prices of outputs and hence the pattern of demand. In short, the projections of demand may well imply a pattern of input and output prices and of production which is different in some significant respects from the patterns embodied in the existing average production function.

Of what use is the existing production function for projecting potential output under these circumstances? The answer depends upon the magnitude of the implied changes in price and production patterns. If the changes are substantial then the existing function will be of very limited usefulness as a projection device. But in a period of 5 years or less – the plan period here under discussion – changes in price and production patterns are unlikely to be so pronounced as to render the existing function useless. The total demand for food, for example, is unlikely to grow by much more than 20 per cent over such a period,[17] nor, with per capita income rising no more than 10–12 per cent, are sharp changes in the pattern of demand likely. With relatively modest changes in both total demand and its pattern, there is little reason to expect such sharp changes in relative prices of inputs and outputs or in the pattern of production as to render useless the existing average production function.

This generalization will of course not be equally valid in all countries, and it may be completely invalid in some. The planner would have to make a judgment in this regard in accordance with the particular situation he faces. It is noteworthy, however, that the generalization is supported by the experience of Chile in the 1950's. Over the course of the decade total agricultural production increased by about 25 per cent, but the pattern of production seems to have been quite stable. This is suggested by the data in Table 46, which show that between the mid-1940's and mid-1950's the only changes of any importance in the relative contribution of various commodities to total output were a drop of 5.6 percentage points in cereals and an increase of 3.7 percentage points in vegetables. With respect to prices, Table 2 indicates that over the decade of the 1950's relative prices of inputs and outputs showed little change, when the input price index includes wages. Excluding wages, the price relationship was quite stable in the first half of the decade but less so in the second half.

The Chilean experience suggests that even in a period of rampant general inflation, agricultural production patterns and relative prices of inputs and outputs may be reasonably stable over periods as long as a decade. For planning periods of no more than 5 years even greater stability could be expected. Thus the planner should not be deterred from using the existing average production function for projections simply because future production

[17]With a 2.0 per cent per annum increase in per capita income, income elasticity of demand for food of 0.8 per cent and population growth of 2.5 per cent annually, total demand for food would rise by about 20 per cent in five years.

Table 46. Percentage Distribution of Production of Agricultural Commodities in Chile, 1945–46–47 and 1955–56–57

	Average for	
Commodity	1945–46–47	1955–56–57
Cereals	21.6%	16.0%
Legumes	4.7	4.0
Tubers	7.9	9.1
Vegetables	3.9	7.6
Fruits	5.7	5.3
Wine	8.5	8.4
Industrial crops	1.9	1.9
Meat	22.3	23.6
Milk products	11.9	10.6
Wool	2.8	3.9
Poultry products	6.4	7.4
Fish	2.4	2.2
Total	100.0	100.0

Source: Presidencia de la República, Oficina de Planificación Nacional, *Indice de producción agropecuaria-silvícola, 1939-1964* (mimeo., not paginated; Santiago, 1966).

patterns and price relations may be different from those embodied in the existing function. The important question is the extent of these differences. There is good reason to believe that they will be within acceptable limits.

There is an additional source of difficulty, however, noted above. The average production function embodies the existing average level of technology. What happens to its predictive value if the level of technology changes significantly? Since one of the principal objectives of agricultural development planning is precisely to promote technological advance, this obviously is a highly relevant question. The immediate response to it is a counter question: How much technological change can occur in 5 years in the agriculture of the typical less developed country? In considering this question it is useful to recall once again that technological change is a process of substituting new kinds of inputs for old. If the existing production function reflects a technology consisting predominantly of a mix of land, unskilled labor, unimproved seed, work animals, and simple plows and other hand tools, then the substitution of skilled for unskilled labor, tractors for animals, and improved for unimproved seed, and the widespread introduction of fertilizers and pesticides, constitutes a change in technology which could not possibly be represented by the old production function. But how likely is it that in a period of but 5 years substitution of this sort will occur on such a scale as to shift the greater part of agriculture from the old to the new function? If the primitive production function is indeed representative, then even if the technical availability of modern inputs and the economic ability of farmers to use them presented no obstacles, wholesale substitution of these inputs would still be seriously impeded by the inability of farmers to manage them properly. To acquire these skills on the necessary scale would require massive programs of general and technical education and extension. As noted previously, such programs take considerable time to mount and yield their fruit but slowly. Five years almost surely would not be enough to do the job, even if the will existed and the necessary resources were available.

Thus the planner need have little concern that in the course of a 5-year planning period a production function embodying primitive technology will be replaced by one embodying modern technology. If the existing function is indeed primitive, then most of whatever output increase occurs over the next 5 years will be achieved with this function. It will therefore serve for projections.

If the existing function is not primitive but already embodies substantial amounts of modern inputs, then the increased quantities of these inputs which can realistically be mobilized over the next 5 years are not likely to change the function significantly. For in this case there is no substitution of modern for primitive inputs. To be sure the new inputs may enjoy some advantages over the old ones, such as a somewhat more nutritive component per pound of fertilizer or a bit more horsepower per tractor. But these are marginal changes compared to the quantum jumps represented by the switch from mules to tractors or from no fertilizer to ample use of it.

Thus if the existing production function represents basically a primitive technology or basically a modern one, the planner can use it with reasonable confidence for projecting potential production over a 5-year planning period. However, it is probable that the more typical situation is one in which the average function embodies both primitive and modern technologies to some extent. In this event the planner must judge the effects on the average function of changes which somewhat increase the importance of modern technologies while diminishing that of the primitive ones. There are a number of elements that ought to be considered in reaching this judgment. In the first place, the planner must ask himself if the average function is really meaningful. If, for example, about half of the agricultural production of the country is generated by farms using essentially a primitive technology while the rest comes from modern farms, then the average production function is a hybrid of doubtful significance. This fact may be revealed by high variances of the regression coefficients in the function, indicating that the underlying data do not yield a tight set of input-output relationships. In this event the best course is to break the function up, deriving separate ones for the primitive and the modern technology farm groups.

Frequently, however, it may be found that the statistical fit of the average production function is quite good — for example, like that of the 177 non-fruit farms discussed in Chapter 4 — even though it is apparent that the function is a hybrid reflecting conditions on farms using both primitive and modern technologies. The good fit, however, will indicate either that the differences between modern and primitive technologies are not extreme or that one or the other is strongly dominant. In either event, the function should serve to yield first approximation projections of potential production. Since a basic objective of planning is to "modernize" the production function, the planner must expect that, whatever the mix of primitive and modern elements in the existing one, the modern will become more important over the plan period. Hence, he will wish to revise the first approximation projections. A crude but simple procedure for doing this would be to assume that the quantity of modern inputs which actually will be used over the plan period will be X per cent greater than the quantity indicated as most economical by the existing production function. The amount of X would depend basically on the planner's assessment of the strength of the institutional obstacles to greater production and incorporation of modern inputs. Of course, the projection of output also would be revised upward in some relation to X. While the specific relation probably would be arbitrary, a general rule would be that the percentage increase in the projection of output would be less than X, since some inputs – land, for instance – would be increased proportionately less than the modern inputs.

These suggestions for modifying projections of potential production and inputs to allow for marginal shifts in the existing average production function lay down no systematic and detailed set of rules. Obviously, ad hoc procedures are called for involving considerable judgment by the planner. The idea underlying the suggestions is generally applicable and sensible, however, namely that some technological improvement will occur over the plan period, implying that both quantities of modern inputs employed and potential production will be somewhat greater than would be indicated by the existing average production function.

The discussion of the last few pages has attempted to show that the existing average production function, perhaps with some minor ad hoc adjustments, can be used to project potential output and corresponding inputs despite the three apparently restrictive assumptions: constant input-output price relations, production patterns, and technology. However, this optimistic conclusion in every case has relied heavily on the assumption of a relatively short planning period – 5 years. As the planning period becomes longer, the conclusion is increasingly undermined, for obvious reasons. The greater the time involved, the more chance everything has to change. The Chilean experience cited earlier in this chapter suggests that even in as much as a decade relative prices and production patterns may be reasonably stable. That is a too slender base on which to generalize, however, and in any case it must be expected that significant changes in technology will occur in 10 years unless the planning effort is completely moribund and the economy generally lacking in dynamism. Whatever the institutional obstacles to the incorporation of more modern inputs, the likelihood of overcoming or weakening them increases with time.

Hence, the further the planning horizon is extended beyond 5 years, the less reliable become the projections of potential output and inputs obtained with the existing average production function. The point at which the unreliability becomes so great as to vitiate use of the procedure will vary from country to country, depending in large part upon the strength of the institutional obstacles to technological change and upon the vigor and skill with which these obstacles are attacked. In general, however, projections much beyond 5 years probably should be taken as no more than rough indicators of the possible.

The function for the top X per cent of farms may yield acceptable longer term projections, however. It is reasonable to suppose that over time average technology will move in the direction of that now employed by the most advanced farmers. Hence the production function for the present top X per cent of farms may provide the planner with a good approximation to the function for all farms 5 to 10 years in the future. This is long enough to permit presently mounted programs of farmer education, training, and extension to have an impact on the average level of skills, which is the most basic factor limiting expansion of output over shorter periods.

POLICY FORMULATION

All too often planning has been identified with the formulation of a set of projections, resulting in plans which contained no blueprints for action, no guidelines for policy makers. The procedures employed in this study would provide powerful assistance to the planner in establishing the relationship between plan targets and the policies necessary for achieving them. Having identified and measured the contribution of the major sources of output growth, the planner has valuable insight into which inputs should be increased, by how much, and what the total impact on production will be. He thus can provide the policy maker with a good indication both of the total volume of resources required to achieve production targets and of the amounts needed of particular inputs. Having analyzed the institutional obstacles to the increased flow of these inputs and formed a judgment of the relative strength of these obstacles over the plan period, the planner can advise the policy maker as to which parts of the institutional structure are most critical and most amenable to change.

The analysis of the Chilean situation illustrates these points. With the results of this analysis in hand the Chilean agricultural planner in the early 1960's would have been in a position to provide the policy maker with valuable information concerning feasible production targets and the policies needed to achieve them. He would have pointed out that the principal obstacles to achieving the potential output attainable over, say, the first half of the 1960's were not unfavorable price relations or the scarcity of irrigated land, or the land tenure system, or the "uneconomic" attitudes of large landowners, but rather credit and foreign-trade policies which prevented farmers from employing more fertilizers, pesticides, and farm machinery. Hence, he would have urged that the plan for the first half of the 1960's ought to incorporate changes in these policies designed to increase the supply of the critical inputs in the amount indicated by the production function. With respect to credit, the planner would have pointed out to the policy maker not only that more should be made available to agriculture, but that the existing system for granting it produced some undesirable results. It placed small farmers at a serious disadvantage in the competition for modern inputs, and it tended to distort the price of land by making that resource the key to credit availability. Thus the planner would have had something to say to the policy maker both about the total amount of credit that ought to be placed at the disposal of agriculture and about desirable changes in the system of credit distribution.

With respect to the longer term, say, the entire decade of the 1960's, the planner could have used the productivity difference between, for example, the 60 most productive nonfruit farms and the whole group (177) of such farms to indicate the scale of feasible increases in output achievable by the end of the decade if the necessary inputs were available. Among these inputs the planner would have to give high importance to improved human skills, since differences in this respect must surely account for some of the observed differences in productivity. Accordingly, he probably would recommend an immediate increase in the flow of resources into both general and technical education for farmers as well as into a considerably expanded extension effort. Because of the probable effects on farmer incentives and on the distribution of power in the countryside, the planner very likely would also recommend a land reform program pointing toward more equal distribution of land.

These examples of how the agricultural planner in Chile could have drawn policy recommendations from this analysis of the Chilean situation are purely illustrative. They are meant only to show how the procedures employed in the analysis produce policy-relevant results. No claim is made that this particular study would have provided the Chilean planner with enough information to draw firmly grounded policy conclusions covering all aspects of the country's agricultural economy. Obviously a study based on only 210 farms from only one agricultural area of the country is not sufficient for that. The claim is made, however, that the procedures employed, if applied to a sufficiently wide range of commodity and regional data and if backed by sufficient research resources, would allow the planner to present the policy maker with a set of projections of output targets, indicating the total amounts of various resources needed to achieve those targets and the policy alternatives most likely to assure that those resources would actually be available to and employed by farmers.

SUMMARY ON PLANNING

The greater part of this chapter has tried to show that the procedures followed in the analysis of Chilean agriculture in the 1950's have considerable relevance to agricultural planners generally. In particular, they provide the planner with a fruitful mode of attack on the three major parts of his assignment: diagnosis, projections, and formulation of policy recommendations. While the emphasis in the discussion has been on the positive values of these procedures in facilitating the planner's work, their various limitations have also been pointed out. This aspect bears reemphasis to avoid any misunderstanding of what the suggested procedures can and cannot do for the planner. They do not provide a panacea, a magical formula for solving all his problems. As noted in connection with demand projections, they do not even touch some important aspects of his work. Another major task to which these procedures do not apply is project formulation. They will tell the planner that at present prices, achievement of output potential will require X more tons of fertilizers, but they will not tell him precisely which kinds of fertilizers (unless the production function is specified in far more detail than those used here) nor whether they should be produced domestically or imported. They may indicate that the return to irrigated land is sufficiently high to justify the addition of Y more acres at present prices, but they will provide no guidance in design of the projects required to irrigate that much more land. The procedures may lead the planner to conclude that substantially more resources ought to go into agricultural research, extension, and farmer education, but they will indicate nothing about where the research laboratories, extension stations, and schools should be located nor what they will cost to construct and operate. All these instances involve detailed analyses of particular projects, a task to which the suggested procedures are not applicable.

The fruitfulness of the procedures in handling those parts of the planner's job to which they are applicable depends crucially upon the quality of the data available, the range of its commodity and regional coverage, and the size, skill, and imagination of the planning effort. The importance of data has been mentioned before. The essential requirement is that they be representative of the full range of the country's agricultural economy. Representativeness, of course, can be achieved through properly designed and conducted sample surveys. Censuses have their value, but they are not essential to implement the procedures suggested here. In general, the greater the number of inputs for which data are collected the better. This permits a more complete specification of production functions, so that the planner is provided with more reliable information concerning the sources of output growth. For any given amount and quality of data the success of the planning effort will depend upon the number and skills of the persons assigned to it. This fact is so obvious as to require little comment. No generally valid statements can be made about the optimum size of the planning effort or about the range of needed skills. Obviously, however, economists, statisticians, computer programmers, sociologists, and political scientists will all have important roles to play if the procedures are to be properly implemented.

No single analytical technique will be adequate to deal with the full range of the planner's task. Like any good workman, he must equip himself with a set of tools, each suited to a particular job. It is believed, however, that the procedures employed in this study, properly understood and used with skill and imagination, constitute a valuable addition to the planner's tool kit. No other claim is made for them.

Summary of 1968
Sample Survey Results *

In March 1968, a survey was taken of 63 farmers in O'Higgins Province. The production and input data collected referred to the crop year 1967/1968. All the farms surveyed were among those included in the 1958/1959 sample analyzed in Chapters 4 and 5, although one of them has been divided into 4 smaller, separately managed farms. While the farms were among those included in the previous survey, there had been rather considerable changes in ownership in the intervening 9 years, and average farm size had been reduced. Two interviewers gathered the information, both of them *ingenieros agrónomos* (holding university degrees in agriculture) with specialties in agricultural economics. Both had had personal experience in farm management, one of them in O'Higgins Province itself and the other in Concepción Province.[1]

The purpose of the 1968 survey was to determine if the principal characteristics of MPF and LPF[2] were the same as those discovered by analysis of the 1959 data. However, circumstances did not permit collection of as much input information as in the earlier survey, so that calculation of production functions for farms in the 1968 survey was not possible. Consequently, the techniques employed on the earlier data to measure differences in produc-

---

*This survey was undertaken with the unstinting cooperation and assistance of personnel of the Agricultural and Livestock Production Division of the Agricultural and Livestock Service, Ministry of Agriculture, both in Santiago and in Rancagua, the regional headquarters for O'Higgins Province. Their generosity is most gratefully acknowledged.

[1]Mr. Alden Gaete and Mr. Atilio Giglio were the principal interviewers. Each gave considerable assistance also in the design of the questionnaire and in tabulation of the results. Mr. Cornelio Marchán also participated at all stages of the survey and made a most valuable contribution.

[2]Most productive farms and least productive farms.

tivity and to isolate the sources of these differences could not be employed in the analysis of the more recent data. Instead, more simple procedures were adopted. Farms were divided into two groups according to the value of gross output per irrigated hectare: an MPF group of 32 farms and an LPF group of 31. Ten tables were then prepared setting forth the principal characteristics of the MPF and LPF thought to be associated with differences in productivity. The tables, which generally speak for themselves, are presented below. The principal results are summarized in the following paragraphs:[3]

PRODUCTIVITY

1. Gross value of output per hectare was over three times greater on MPF than on LPF. Per man-day worked the MPF produced 2.6 times as much as the LPF.[4]

2. A large part of these differences was due to the fact that MPF farmed a greater proportion of their land than LPF; a smaller part reflected somewhat higher prices received by MPF than by LPF.

3. However, the physical volume of production per hectare actually farmed was 55 to 65 per cent greater on MPF than on LPF.

4. The relative contributions to production of various commodities were about the same for the two groups of farms.

For further information, see Tables A-1, A-2, and A-3.

QUANTITY AND QUALITY OF LAND

1. MPF were larger than LPF, in both irrigated and total area. The differences were relatively small, however. The ratio of irrigated to total farm area was slightly less on MPF.

2. With respect to fertility, drainage, and absence of rocks, the farmers in the MPF group rated 74 per cent of their land as being either good or very good, while the corresponding percentage for LPF was 48 per cent.

---

[3] Most of the material presented in the tables is in the form of ratios and averages summarizing principal characteristics of the LPF and the MPF. No attempt has been made to apply statistical tests of significance to the differences between these various figures. Nevertheless, as noted in the summary preceding the tables, the pattern which emerges from the figures is plausible and, we believe, significant.

[4] A better measure of productivity is the amount of production per unit cost of *all* resources employed. Conceivably, a group of farms could obtain higher yields per hectare and per man than another group but still have lower yields per unit of total cost. This could happen if the higher yields were obtained on the first group of farms by employment of much greater quantities of nonland, nonlabor inputs than on the second group of farms. The higher cost of these inputs possibly could more than offset the lower costs of land and labor per unit of output.

The 1968 survey did not yield sufficient information to measure total costs of production. However, the data collected on costs of fertilizers, seeds, and pesticides, the amounts of land, labor, and numbers of tractors employed, combined with general information concerning wage rates and reasonable assumptions about depreciation and all other costs, leaves no doubt that production per unit cost of all resources on the MPF was substantially higher than on the LPF.

3. Consistent with this finding, the median value of land per hectare, as estimated by the farmers, was 26 per cent higher on MPF than on LPF.

4. Among farmers who said they would like to buy more land, the MPF indicated they would be willing to pay about 10 per cent more per hectare than the LPF.

5. Comparing the amounts farmers were prepared to pay to rent more land with their estimates of the value of land suggests that the rate of return to land investments was about 10–12 per cent.

For further information, see Table A-4.

FERTILIZER

1. Per hectare of irrigated land fertilized the MPF employed 17 per cent more nitrate fertilizer, 124 per cent more potassium fertilizer, and 14 per cent more super triple phosphate fertilizer than the LPF. The total quantity of fertilizer used per hectare of all irrigated land was 233 per cent greater on MPF than on LPF.

2. Fifty-three per cent of those MPF which fertilized used machinery to do so compared with only 20 per cent of the LPF. Ninety-four per cent of the MPF applied some amount of fertilizer, but only 81 per cent of the LPF did so.

For further information, see Table A-5.

PESTICIDES

1. Seventy-five per cent of the MPF applied some amount of pesticides compared with only 52 per cent of the LPF.

2. The LPF spent 18 per cent more on pesticides per hectare of land to which pesticides were applied. However, because they applied pesticides to a greater proportion of total land area, the MPF spent 53 per cent more per hectare of all irrigated land than the LPF.

For further information, see Table A-6.

SEEDS

1. Over 54 per cent of total seed expenditures by MPF were for improved seed varieties compared with 49 per cent for LPF.

2. On the MPF 80 per cent of the land sown to annual crops was in improved seed varieties while for LPF the corresponding figure was 72 per cent.

3. Per hectare of land sown to improved seed varieties, seed expenditures were 40 per cent greater on MPF than on LPF. Per hectare of land sown to unimproved seed varieties, the MPF spent 82 per cent more than the LPF.

4. Total seed expenditures per irrigated hectare were 67 per cent greater on MPF. Per man-day worked the MPF spent 33 per cent more on seeds than the LPF.

For further information, see Table A-7.

TRACTORS

1. Eighty-one per cent of the MPF had at least one tractor compared with 58 per cent of the LPF.

2. The average MPF tractor had 48 horsepower and was 6.6 years old. The LPF tractors averaged 44 horsepower and 7.3 years.

3. The MPF had 37 per cent more horsepower per irrigated hectare than the LPF.

For further information, see Table A-8.

LABOR

1. The MPF employed 33 per cent more labor per farm than the LPF and 25 per cent more per irrigated hectare.

2. Forty-four per cent of the labor input on the LPF was by *inquilinos* (resident labor), the traditional mode of contracting farm labor in Chile. *Inquilinos* provided 38 per cent of the labor on the MPF. A little over 50 per cent of labor inputs on MPF were nonresident laborers compared with 45 per cent on the LPF. The MPF also employed a slightly higher percentage of specialized labor and a somewhat lower percentage of family labor than the LPF.

For further information, see Table A-9.

MANAGEMENT

1. Fully two-thirds of the least productive farmers said that their farming operations were sufficient to absorb all their time. By contrast, only one-third of the most productive farmers were thus fully occupied. Most of those not engaged full-time on their own farms worked on other properties, although several of the MPF indicated they had machine rental or other businesses as sideline activities.

2. The average MPF was 46.7 years old and had been a farmer for 21.3 years. The LPF averaged 52.0 years of age and 22.6 years in farming.

3. Twenty per cent of the LPF had three years or less of schooling (10 per cent had none) compared with 6 per cent for the MPF.

4. Fifty-two per cent of the MPF had completed 12 years of school while only 37 per cent of the LPF had done so.

5. Thirty per cent of the MPF had completed one or more years of university work or had received special technical training in agriculture compared with 10 per cent of the LPF.

6. Ninety-one per cent of the MPF indicated that they had introduced one or more changes in production techniques in the previous five years while only 74 per cent of the LPF had done so.

7. Forty-eight and one-half per cent of the MPF listened regularly or occasionally to a radio program dealing with agriculture compared with 58

per cent of the LPF. However, 79 per cent of the MPF indicated that in the previous year they had read one or more articles dealing with agriculture while only 55 per cent of the LPF had done so.

For further information, see Table A-10.

QUANTITIES OF INPUTS PER UNIT OF LAND AND LABOR

1. Per hectare of irrigated land the MPF employed 26 per cent more labor; spent 233 per cent more on fertilizer, 53 per cent more on pesticides, and 67 per cent more on seeds; and had 37 per cent more tractor horsepower than LPF.

2. Per man-day worked the MPF had only 80 per cent as much irrigated land as LPF. However, the MPF had almost 14 per cent more land in crops per man-day worked than did the LPF. The differences reflect the greater percentage of MPF irrigated land in crops.[5]

3. Per man-day worked the MPF spent 69 per cent more on fertilizer than the LPF and 33 per cent more on pesticides and seeds, and had 9 per cent more tractor horsepower.

For further information, see Tables A-2 through A-9.

APPENDIX TABLES

Table A-1. Various Output Relationships

	32 MPF	31 LPF
Total production	E°7,167,159	E°1,935,574
Production per hectare:[a]		
Total area	E°1,822	E°579
Irrigated area	E°2,628	E°800
Production per man-day worked	E°69.5	E°26.6
Per cent distribution of production:		
Animal products	15.3%	14.5%
Crops	84.7	85.5
Corn	23.8	23.2
Wheat	16.7	16.8
Potatoes	11.3	4.6
Beans	8.7	11.5
Fruit	6.1	6.3
Clover	1.6	1.6
Alfalfa	1.3	6.2
Other crops	15.2	15.3
Per cent of production –		
Sold	97.0	94.8
Consumed on farm[b]	3.0	5.2

[a]1 hectare = 2.47 acres.
[b]Includes feed.

[5] Recall (from Table A-1) that on both MPF and LPF crops accounted for about 85 per cent of the total value of gross output.

Table A-2. Output per Hectare

	Indexes (MPF ÷ LPF × 100)
By value, all products:	
Per total hectares	315
Per irrigated hectare	329
By quantity, 8 major crops:	
Weighted by MPF shares	165
Weighted by LPF shares	156

	Quintals per hectare[a]	
	MPF	LPF
By quantity:		
Wheat	46.1	30.9
Corn	72.9	47.6
Potatoes	205.4	111.0
Beans	27.8	15.9

Notes:

Indexes by value were derived from Table A-1. Indexes by quantity are averages of ratios of physical quantities of output per hectare of the 8 major crops weighted by share of each crop in total production. These 8 commodities accounted for 87.0 per cent of total crop production of MPF and 86.6 per cent of that of LPF.

The differences between the value and quantity indexes reflect both higher prices received by MPF and the greater proportion of all irrigated land devoted to crop production on MPF. (See Tables A-3 and A-4.)

[a] 1 quintal = 220.5 lbs. According to the 1965 agricultural census, yields in O'Higgins Province were 28.9 quintals, 40.8 quintals, and 133.1 quintals for wheat, corn, and potatoes, respectively. Allowing for some increase in productivity between 1965 and 1968, for the fact that 1965 was a very dry year (ODEPA figures show that total crop production in the country fell 7.5 per cent from 1964 to 1965), and for the probability that the farms in our sample, being relatively large, were more productive than the province average, the sample yield figures do not appear excessively high relative to the province averages for 1965.

Table A-3. Average Prices Received[a]

	Escudos per quintal		
	MPF	LPF	MPF ÷ LPF
Wheat	44.8	40.4	1.109
Corn	40.6	35.0	1.160
Beans	85.4	74.9	1.140
Potatoes	28.7	17.1	1.678
Alfalfa	20.5	14.2	1.444
Rice	58.5	40.6	1.441
Barley	31.9	33.4	0.955
Sunflower seed	101.8	56.6	1.799

[a] Total value of production divided by total quantity of production in quintals.

Table A-4. Various Land Relationships

	MPF	LPF
Average amounts of land per farm (hectares):		
Total	122.9	107.9
Irrigated	85.2	78.0
Per cent irrigated	69.3	72.2
Per cent of farms with less than –		
10 irrigated hectares	9.4%	6.5%
20 irrigated hectares	22.5	22.6
50 irrigated hectares	37.5	51.6
100 irrigated hectares	62.5	74.2
200 irrigated hectares	87.5	90.3
305 irrigated hectares	100.0	100.0
Man-days worked per hectare of irrigated land	37.8	30.1
Per cent of irrigated land which, with respect to fertility, drainage, and absence of rocks, is –		
Very good	17.0%	11.0%
Good	56.9	37.1
Average	20.4	35.1
Bad	5.7	16.8
Per cent of irrigated land in crops	78.3	58.5
Per cent of farms whose soils have been analyzed	46.9	32.3
Per cent of land which is –		
Owner operated	86.5	69.8 (89.4)[a]
Rented	13.5	30.2 (10.6)[a]
Land values per irrigated hectare (estimated by farmer):		
Mean	E°4,024	E°3,433
Median	E°4,000	E°3,167
Per cent of farmers who would like to buy or rent more land	53.1%	51.6%
Price those willing to buy or rent would be willing to pay (per hectare)		
If buying:		
Mean	E°3,792	E°3,400
Median	E°3,333	E°3,000
If renting (per year):		
Mean	E°425	E°442
Median	E°382	E°332
Implicit rate of return to land (per cent) if price were –		
Equal to that farmers were willing to pay		
Mean	11.2%	13.0%
Median	11.5	11.1
Equal to that placed by farmers on own land		
Mean	10.6	12.9
Median	9.6	10.5

[a]Excludes one farm with 200 owned and 703 rented hectares.

Table A-5. Various Fertilizer Relationships

	MPF	LPF
Amount of fertilizer applied (quintals) per hectare — Of irrigated land fertilized:		
Nitrates	4.1	3.5
Potassium	6.5	2.9
Super triple phosphate	2.5	2.2
Of irrigated land:		
All fertilizers	5.0	1.5
Amount of all fertilizers (quintals) per man-day worked	.132	.078
Per cent of farms applying fertilizer	93.8%	80.7%
Per cent of fertilizer-using farms which used machinery to fertilize	53.3	20.0
Per cent of fertilizer-using farms which purchased fertilizer with credit	76.7	75.0

Table A-6. Various Pesticide Relationships

	MPF	LPF
Per cent of farms applying pesticides	75.0%	51.6%
Expenditures for pesticides per hectare —		
Of land to which pesticides were applied	E°89.1	E°105.5
Of all irrigated land	E°40.5	E°26.5
Expenditures for pesticides per man-day worked	E°1.12	E°0.84

Table A-7. Various Seed Relationships

	Per cent of total seed expenditures represented by —	
	Improved seed varieties	Unimproved seed varieties
Most productive farms	54.4%	45.6%
Least productive farms	48.9	51.1

	Per cent of hectares sown to —			
	Improved seed varieties		Unimproved seed varieties	
	MPF	LPF	MPF	LPF
Wheat	100.0%	89.7%	0.0%	10.3%
Corn	96.8	81.8	3.2	18.2
Beans	49.6	0.4	50.4	99.6
Potatoes	0.0	0.0	100.0	100.0
Alfalfa	56.9	86.2	43.1	13.8
Clover	98.2	90.2	1.8	9.8
Rice	100.0	100.0	0.0	0.0
All other annual crops	41.2	68.2	58.8	31.8
All annual crops	80.0	71.7	20.0	28.3

Table A-7. *Continued*

| | Expenditures (escudos) per hectare sown to − | | | |
| | Improved seed varieties | | Unimproved seed varieties | |
	MPF	LPF	MPF	LPF
Wheat	E°151.4	E°134.0	0.0	E°100.5
Corn	104.6	84.7	E°48.9	68.1
Beans	480.3	437.5	524.2	326.5
Potatoes	0.0	0.0	998.0	916.3
Alfalfa	43.4	118.6	n.a.	n.a.
Clover	75.1	70.1	88.0	66.8
Rice	195.2	54.0	0.0	0.0
All other annual crops	17.8	65.2	101.6	106.0
All annual crops	145.2	102.8	494.0	271.7

| | Total seed expenditures | |
	MPF	LPF
Per irrigated hectare	E°166.0	E°99.3
Per man-day worked	E°4.4	E°3.3

n.a. Not available.

Table A-8. Various Tractor Relationships

	MPF	LPF
Per cent of farms with tractors	81.3%	58.1%
Total number of tractors	52	39
Average horsepower	48[a]	44[b]
Average age in years	6.6	7.3
Horsepower per irrigated hectare	.82	.60
Horsepower per man-day worked X 100	2.16	1.99

[a]Average for 46 tractors. Information not available for 6.
[b]Average for 33 tractors. Information not available for 6.

Table A-9. Various Labor Relationships

	MPF	LPF
Man-days worked per farm	3,129	2,346
Man-days worked per hectare of –		
Irrigated land	37.8	30.1
All land	26.2	21.7
Per cent of total man-days worked which were –		
Family labor	4.4%	5.3%
Specialized labor	7.4	5.9
Unspecialized resident labor	37.9	43.9
Unspecialized nonresident labor	50.3	44.9
	100.0	100.0
Per cent of farmers' time spent in –		
Working directly in the fields	13.8%	11.4%
Supervising work of employees	57.7	69.1
Other management-related work	28.5	19.5
	100.0	100.0

Table A-10. Various Management Relationships

	MPF	LPF
Average age of farmers	46.7	52.0
Average number of years as farmers	21.3	22.6
Per cent of farmers–		
With no schooling	3.0%	10.0%
With 1–3 years schooling	3.0	10.0
With 4–6 years schooling	15.2	3.3
With 7–9 years schooling	9.1	20.0
With 10–12 years schooling	69.7	56.7
With 12 complete years schooling	51.5	36.7
With 1 or more years of university, or specialized		
training in agriculture	30.3	10.0
Per cent of farmers indicating that work on the farm was		
sufficient to absorb 100 per cent of their time	33.3	67.7
Per cent of farmers who listen to radio programs		
dealing with agriculture	48.5	58.1
Per cent of farmers who in previous year had read		
material dealing with agriculture	78.8	54.8
Per cent of farmers who had changed production		
techniques within previous five years	90.6	74.2
Per cent of farmers who, compared with five years ago–		
Had substituted tractor for animal power	53.1	58.1
Used more fertilizer	78.1	58.1
Used more improved seed varieties	59.4	61.6
Used more pesticides	71.9	61.6
Had changed irrigation techniques	34.4	22.6

DISCUSSION OF RESULTS

The pattern of differences between the MPF and LPF is very similar in all important respects to that found in Chapter 5, even though the measure of productivity in the present case is output per hectare whereas in the earlier case it was total productivity. The upper 50 per cent of farms were substantially more productive than the lower 50 per cent and were larger, whether size is measured by land area or total inputs. The MPF employed more of all nonland inputs per hectare of irrigated land, the largest differences in descending order being in fertilizers, seeds, pesticides, tractor horsepower, and labor.

The differences that exist in characteristics are mostly matters of degree. For example, differences in average prices received were more pronounced (in favor of the MPF) in the 1968 sample than in that taken in 1959, an indication that prices were responsible for a greater proportion of productivity differences in the later than in the earlier sample. Also, in the 1968 sample the proportion of irrigated land in crops was higher for the MPF than for the LPF, whereas in the 1959 data this proportion was about the same for both groups of farms. This indicates a more intensive use of land by the MPF in the later sample and hence accounts for some of the difference in productivity. In the earlier sample, intensity of land use was not a factor in productivity differences. Unfortunately, no information is available indicating why MPF in the 1968 survey received higher prices for their products and utilized their land more intensively than LPF. Better knowledge of product markets – knowing when and where to sell to best advantage – may partly explain the higher prices received, as may quality differences in products. Obviously there may be fairly wide quality differences in such commodities as corn, wheat, and potatoes, depending upon the particular varieties grown and upon the care and skill with which they have been planted, cultivated, harvested, and stored.

It was concluded in Chapter 5 that managerial differences were secondary to machine economies in explaining the total productivity differences. However, it was noted that the 1959 data provided little direct evidence relating to management. The 1968 survey data are more helpful in this respect, and they suggest that management was significant in explaining productivity differences. The more productive farmers had considerably more education than the less productive farmers and they were more inclined to read material on agriculture and to adopt technical innovations. While the quantitative impact of these factors on productivity cannot be determined from the survey data, there is every reason to believe it was important.

The significance of the finding that only one-third of the more productive farmers found the work of their farms sufficient to absorb all their time, compared to two-thirds of the LPF, is not obvious. The question was included in the survey to provide information relevant to the hypothesis that full-time farmers are better at their jobs than part-time farmers. At first glance the results seem to contradict the hypothesis. However, the contradiction is less marked when one notes that 12 of the 22 MPF not fully occupied on their own farms spent the balance of their time working on other farms, or engaged in business related to agriculture. Of the 10 LPF not occupied full-time on their own farms, only 3 found employment elsewhere in agriculture.

The difference in percentages of farmers fully occupied on their own farms may be related to managerial differences, but in a way opposite to that suggested by the hypothesis mentioned above. More energetic and enterprising farmers are likely to be better managers than those less well equipped in these respects. Hence, the greater percentage of MPF not fully occupied on their own farms may reflect greater energy and enterprise among these farmers than among the LPF. This hypothesis gains considerable support from the fact that the 22 MPF *not* fully occupied on their own farms had an average of 61.1 hectares in crops, while the 21 LPF who *were* fully occupied on their own farms had only 39.1 hectares in crops. This difference strongly suggests that the MPF were simply able to handle more work than the LPF.

The data in Table A-4 on farmers' estimates of the value of land and the amounts they would be willing to pay to rent an additional hectare suggest that the marginal return to land was some 10–12 per cent. While this is below the opportunity cost of capital (15 per cent) assumed in Chapter 4, it is considerably above the estimated marginal productivity of land (6 per cent) derived from the production function for the 1959 sample of 177 nonfruit farms. Is it reasonable to expect that the return to land would have risen in this period, or are we dealing simply with sampling fluctuations in the data?

There are two factors bearing on changes in the marginal return to land. One is relative prices of land and farm commodities, and the other is the marginal physical productivity of land. Assuming the latter to be constant, then the return to land will rise if farm commodity prices increase faster than the price of land. While the data leave much to be desired, they suggest that farm commodity and land prices in O'Higgins Province increased by about the same relative amounts between 1959 and 1968. Comparing prices received by farmers in the 1959 and 1968 surveys indicates increases of about 6.4 times for wheat, 7.3 times for corn, and 6.9 times for potatoes. If (judging from Table A-4) irrigated land in the province was worth about E°3,600 per hectare in 1968, then land prices must have increased approximately 7 times from their 1959 level.[6] As far as they go, therefore, these data do not suggest that the return to land rose because of favorable relative movements in prices of land and farm commodities.

[6]In Chapter 4 a figure of E°460 was taken as the price per hectare of irrigated land in the province. However, an alternative estimate of E°570 also was noted. If the actual price lay between these two estimates, then land values in the province rose approximately 7 times between 1959 and 1968.

There is every reason to believe, however, that the physical marginal productivity of land increased from 1959 to 1968, owing to marked increases in the amounts of machinery and fertilizers employed per unit of land. While data on the employment of these inputs in O'Higgins Province in 1959 and 1968 are not available, there was a sharp increase in the country as a whole in the use of fertilizer and farm machinery in the period from 1959 to 1965.[7] It is highly likely that O'Higgins, one of the most productive and progressive agricultural regions in Chile, shared fully in the greater use of these inputs. Since there was little, if any, addition to the irrigated area of the province, the amount of these inputs per unit of land must have risen rather substantially, with a consequent favorable impact on the marginal physical productivity of land. Of course, there is no certainty that this is the full explanation of the apparent rise in the rate of return to land investments between 1959 and 1968, but given the greater employment of machinery and fertilizer, some increase in the productivity of land would be expected.[8]

An implication of this conclusion is that yields per hectare increased between 1959 and 1968. This in fact appears to have happened. The 63 farms included in the 1968 survey produced an average of $E^o1,768$ per hectare. In 1959 these same farms (but with some differences in ownership and in average size) produced E^o171 per hectare. Of course, the difference between these two figures reflects not only changes in yields but also increases in farm commodity prices. While the price data are quite incomplete, those cited above for wheat, corn, and potatoes suggest that farm commodity prices rose about 7 times between 1959 and 1968. If this were correct, then real output per hectare would have risen by about 48 per cent over this period, an annual average rate of increase of about 4.5 per cent.[9] If prices had risen by a factor of 8 rather than 7, the increase in yields would have been 29 per cent, or almost 3.0 per cent per annum. Hence, these data indicate that the farms included in the two surveys achieved rather impressive gains in real output per hectare between 1959 and 1968.[10]

[7]*El uso de fertilizantes* (see Ch. 2, n. 53), p. 5, indicates that fertilizer consumption in Chile rose by 128 per cent from 1959 to 1965. According to *El uso de maquinaria*, p. 2, (see Ch. 2, n. 4), the stock of farm machinery increased by 27 per cent in this period.

[8]The effect of the 1967 land reform law on land prices in O'Higgins Province is unknown, but they may have been depressed somewhat by the uncertainty surrounding operation of the law. This, of course, would tend to increase the rate of return to land, given its marginal physical productivity and commodity prices.

[9]The relative increase in real output per hectare is equal to the relative increase in the value of output per hectare divided by the relative increase in prices of farm commodities.

[10]This apparently was true also in the nation as a whole. Between 1959 and 1967 national average wheat yields increased 3.7 per cent annually, and for corn the increase from 1959 to 1966 was an astonishing 8.6 per cent per annum (U.S. Department of Commerce, *Agricultural Statistics*, 1961 and 1968). Corn and wheat accounted for about 40 per cent of the total value of output of the 1968 sample of farms.

SUMMARY

Despite some differences in detail, the results of the 1968 survey are quite similar to those derived from analysis of the data for 1959. In both cases, the MPF were found to use more of all inputs, but especially the so-called "modern" ones — fertilizers, pesticides, machinery, and improved seeds — than the LPF. Unfortunately, the information gained in the later survey was insufficient to measure the separate contribution of these various inputs to the observed productivity difference. Nor was it possible to explore the hypothesis advanced in Chapter 5 that this difference was due in large part to economies enjoyed by the MPF in the use of machinery.

Indeed, the 1968 survey results tell little about the causes of differences in productivity. It is superficial to attribute these differences to the greater quantities of "modern" inputs employed by the MPF. One would like to know *why* the MPF were able and willing to make greater use of these inputs. To what extent was it simply because they were better farmers? What role was played by credit and other institutional factors determining access to these inputs? One also would like to know why MPF received higher prices for their products and farmed a higher proportion of their land, both important factors in the observed productivity difference.

The search for answers to these questions would require a separate study of considerable scale, comparable to that focused on the 1959 survey data. Such a study is beyond our means. Consequently, it must be emphasized that the results of the 1968 survey go only part way toward a complete analysis of productivity differences. However, the survey was not undertaken with such an analysis in mind. Its objective was to compare the characteristics of the MPF and LPF in 1968 with those revealed by the 1959 survey. The pattern of productivity differences and input use which emerges is plausible and generally consistent with that found in the 1959 data. A more complete analysis must await more information and another occasion.

Index

Agrarian reform agency, 16
Agricultural plan (1965–80), 10, 34
Ahumada, Jorge, 148, 150
Alcazar, Jorge, xiii
Aldunate, Paul, xiii, 85, 95
Asemblea de Directivas Agrícolas, Comisiones de Politica Agraria y de Problemas Sociales, 97

Balance of payments
and livestock production, 49
problem, 18
Banco Central de Chile. See Central Bank
Banco Interamericano de Desarrollo, 26
Barraclough, Solon, xiii
Barton, G. T., 128
Becker, Gary, 3
Board of Trade (CONDECOR), 19
Bowman, Mary Jane, 3
Bravo, Julio, xiii
Bray, James O., 51
Brubaker, Sterling, xiii
Bruton, Henry, xiii
Burgos, Hernan, xiii

Capital
cost of, 85
fixed: increase rate, 61–62; input measurement, 74; as productivity factor, 130, 141, 142, 143; returns to, 62(t), 63–65, 95; stock of, 54–56(t), 60, 65; utilization, 40 formation, 48
operating, 130, 142, 143
Carmona, César, 41
Carvallo, Marcello, xiii
Catholic University of Chile, 32: Department of Agricultural Economics, xii

Central Bank, 11, 14, 19, 25
Agricultural Committee, 15
credit policies, 99
Executive Committee, 15, 16
Chateauneuf, Rolando, xiii
Chilean Agricultural Planning Office.
See Oficina de Planificación Agrícola, ODEPA
Chilean land reform agency, 11
CIDA. See Comité Interamericano del Desarrollo Agrícolo
Climate, 68, 92
Comité Interamericano del Desarrollo Agrícolo (CIDA), 23, 34–43 passim, 97, 98, 162
Commodities
demand, 7–8
prices: 9(t), 188, 189; indexes, 10(t)
production: MPF and LPF, 181(t); O'Higgins and share of national, 67(t); percentage distribution, 169(t)
Conclusions and implications
capital, 61–62
credit, 145, 146
economies of scale, 145–46
education-technical innovation relation, 146–47
farm size, 137, 145–46, 147
fertilizers, 60, 61, 144–45
general, on inputs, output, and productivity, 60–65
inputs: factor proportions, 127, 130; supply, 102–3; use, 60–65, 86, 89–97 passim, 101, 102, 127, 130, 142–43, 144–45
irrigated land, 86, 95, 96, 142, 144

191